P9-CQJ-576

FIRE SERVICE ADMINISTRATION

FIRE SERVICE ADMINISTRATION

Nancy K. Grant, Ph.D., and
David H. Hoover, Ph.D.

National Fire Protection Association
Quincy, Massachusetts

Project Manager: Deborah Shaw
Project Editors: Cassandra S. Goldwater and Kimberley Foster
Composition: Louise Grant, Marilyn Lupo
Page Layout: W. P. Ellis
Illustrations: Terry LeBlanc
Cover Design: Boston Marketing Services Group
Interior Design: Joyce C. Weston

NFPA No. FSA-94
ISBN 0-87765-390-9
Library of Congress Catalog Card Number: 93-086545
Printed in the United States of America

99 98 97 96 95 94 5 4 3 2 1

Contents

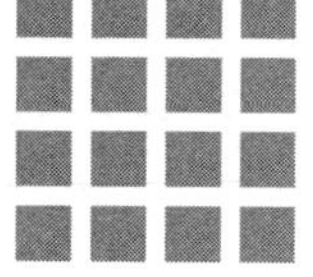

CHAPTER 1

Introduction

The fire service is a respected part of American life, a service begun by volunteers that has developed into a profession responsible for saving lives and protecting property from destructive fire. It is a public organization that takes great pride in its traditional structure and approach to organizational leadership. For decades, the fire service has successfully met the challenge of protecting citizens from fire. Whether serving as a volunteer fire fighter, a paid on-call fire fighter, or a career fire fighter, the opportunity to serve local citizens as a public servant is the ultimate personal reward of a challenging and often dangerous occupation.

The goal of this text is to prepare fire officers to be fire administrators. The book integrates the expectations of fire chiefs and of public administrators with case studies to present various skills, techniques, and theories relevant to administration in the public sector. *Fire Service Administration* focuses on fire service officers to provide assistance in the development of future leadership. The reader can explore topics as necessary or as interest dictates. Further advanced knowledge about a specific field can be researched by using the resources referenced in each chapter. This is not a book to be read and shelved; it can be referenced on an ongoing basis throughout the development of a fire officer's career.

ADMINISTRATION TOPICS MOST RELEVANT TO FIRE SERVICE OFFICERS

Field research was used to identify areas of skill and knowledge necessary for fire administrators. The need for fire administration to more closely parallel public administration is a growing concern of local governments. *Fire Service Administration* provides a framework for developing administrative skills for management in the fire service.

Much of the traditional administration literature and theories can be adapted to the fire service. This information must be tailored to the particular needs of the fire service, however, because the structure and operation of fire departments differ from most other public or private organizations. As an emergency service, staffing levels, preparedness, and daily operational activities must be more flexible in terms of schedule and yet more rigorous in terms of procedures. The daily use of specific rescue techniques and specialized equipment requires continuing technical training. As a result, most fire service training is of a technical nature, very applied and specific.

Management and administrative education, which require critical thinking and applying concepts rather than techniques, can be challenging for members of the fire service. Given the importance of this information, however, this text addresses the challenge by presenting administrative concepts, providing examples of their application to the fire service, and offering various case studies illustrating the management concepts and techniques used in the fire service.

The field of administration is broad. It includes such diverse areas as budgeting, public relations, policy research, and leadership. Some of these areas are more relevant to the duties of the fire officer than are others. As the contents of *Fire Service Administration* were being developed, the authors identified the most important areas of knowledge and expertise by relying on the expert knowledge of practitioners. Fire chiefs and local government officials were identified as the two groups most familiar with the administrative skills necessary, and members of each group were interviewed for this book.[1]

Fire chiefs are familiar with the challenges faced on a daily basis. The public administrators selected were those to whom fire chiefs report. Thus, this set of interviews included the perceptions of professional administrators who have observed the strengths and weaknesses of the administrative preparation of fire officers in their local areas. This provided a different viewpoint of the administrative training needed in the fire service.

These two sets of interviews were conducted around the country. Generally, the fire chief was the officer interviewed from the first group, although on occasion a different administrative officer was identified by either the chief or a public administrator as the individual who would be best able to assist us. The group of local government officials to whom the fire chiefs reported included city managers, mayors, public safety directors, and township trustees.[2]

The interview sample was selected on the basis of regional representation and type of government. Interviews were conducted in the Southwest (Texas), Midwest (Ohio, Indiana, Illinois), East (Massachusetts, Connecticut, New York, Pennsylvania), West (California, Oregon), and South (Florida). In addition, care was taken to ensure a representation of various types of local government: mayor-council, council-manager, commission, and township trustee. Smaller cities (population less than 200,000) in targeted geographic areas were identified and asked to participate. In addition, five cities with populations over 250,000 were included in the interview process. These were Fort Worth, Texas; Akron, Ohio; Chicago, Illinois; Harrisburg, Pennsylvania; and Prince George's County, Maryland. The interviews were conducted over a period of 12 months and involved considerable travel. In all, 35 communities were included in the interview process.

Participation was totally voluntary. Appointments with public administrators were scheduled in advance. During this discussion, the purpose of the interview and the intended use of the information were explained.

The field research utilized "focused interview" techniques. This is a qualitative research methodology used for exploratory research and theory building, or heuristic reasoning endeavors. It was not intended to serve as an evaluative research methodology.

Dr. Richard M. Patterson, manager of staff and auxiliary services with the Kent State University Police Department, reviewed and validated the focused interview process. Dr. Patterson has conducted and published several research studies that utilized the focus group technique as the primary methodology.[3]

The focused interview guide is constant for each interview and yet allows flexibility and individuality in response. The focused interview guide was prepared to ensure that each interview would gather the same information—that the same questions were used to elicit input. (See Focus Interview Form on next page.) This guide was used in each interview, regardless of setting. The authors took notes on the form to facilitate organizing the information gathered in a uniform manner. These forms then provided the database necessary to identify topic areas and specific points to reference in the book.

Each person interviewed was asked to identify those administrative skills and knowledge areas of greatest importance to the fire service. Many interviewees mentioned the same knowledge areas.

As a secondary check to ensure the accuracy of the administrative dimensions identified, the 1992 edition of the National Fire Protection Association's standard, NFPA 1021, *Standard for Fire Officer Professional Qualifications,* was also used to identify administrative and leadership skills important for fire officers. Thus, three sources were used to pinpoint the specific areas of information to be presented in this book: (1) interviews with fire officers, (2) interviews with public administrators, and (3) NFPA 1021, *Standard for Fire Officer Professional Qualifications.* Together they provide a solid basis of administrative information that will enable future fire officers to serve the role of administrator well.[4]

Most of the administrative or management theory available is focused primarily toward managers in private business and industry or toward public administrators. Very little is designed specifically for the fire service. Although there are a number of management-based textbooks, most of them present techniques without detailing the concepts and theoretical background that drive the techniques to work when applied correctly. As a result, it is not always easy for fire service officers to apply the techniques correctly in a variety of situations.

In the same way that technical training contains an element of theoretical grounding, management education must discuss theory as well as application of skills. For example, technical training provides the pump operator/engineer with the knowledge needed to correctly charge a 1¾-in. hand-held hose line to be used for an interior attack or a 4-inch supply line. Although these pressures are different, the pump operator understands the basics of hydraulic water movement and can calculate

FIRE ADMINISTRATION FOCUSED INTERVIEW FORM

Community: ____

Person interviewed: _____________________________________

I. What are your perceptions of the fire service of today?

Prompts: Daily operations, organizational structure, accountability.

II. Where will the fire service be in 10 years?

Prompts: structure, accountability, human resources.

A. Organizationally

B. Technically

III. How do you think they will get there?

Prompts: motivating factors, procedures.

IV. What are the challenges fire departments must face in this process? (resistance)?

Prompts: operations, opportunities, fiscal.

V. What management skills do you think fire service managers will need?

VI. What are your thoughts on mentoring as a viable method of achieving the needed skills?

Who would mentor?

VII. Have you ever discussed any of these issues with your fire chief/city administrator? What was their reaction?

different pressures according to the size of the hose line and the situation in which it is used. Similarly, a fire officer must be able to recognize what type of leadership style is most appropriate for different command situations and use it accordingly. In order to do this, the officer must have knowledge of leadership theory. This is the main contribution of this text. It presents the theoretical background behind managerial concepts so that fire officers can relate them to a variety of administrative settings. Managers and administrators must be able to examine the situation and select the most appropriate administrative technique to be used. Thus, not only will it assist fire service officers in a variety of departments, but as individual departments and the fire service in general change over time, this book will continue to assist officers in dealing with daily management challenges.

HOW THE FIRE SERVICE WILL EVOLVE IN THE COMING DECADE

The fire service in the U.S. is one of the last great bastions of resistance to change. Fire departments excel in the traditional use of technical skills and knowledge to provide a visible and valued service to communities. However, this same tradition has caused fire service administration to remain cloistered, resistant to change, and suspect of outside influences.

In the past, the only measure of a well-managed fire department was that it succeeded in putting out most fires in a reasonable amount of time and provided the citizens with the most appropriate level of fire protection based on the community's resources. Today, a well-managed fire department must be prepared to accomplish much more. Fire administrators are public service department heads. As such, they must be prepared to guide the organization wisely by employing administrative skills and knowledge that are often gained from sources other than technically based fire officer training programs. Fire fighting is no longer the only criterion by which departments are judged effective.

Two national trends are having and will continue to have significant impact on the fire service over the coming decade. The first is the long-term budget deficit facing all levels of government. Services must be provided in a cost-effective manner. Although technical and personnel training advances have been made, there has been little change in the daily operations and duty assignments of fire fighters. Fire officers must now take a serious look at the way the fire service operates and become entrepreneurial in their approach to the service delivery systems of the future. This means that innovative approaches to managing must be initiated in order to provide quality service in new and dynamic ways while simultaneously cutting the cost of these services.

The second trend is the shift in the nature of fire department responsibilities. Emergency Medical Services (EMS) were adopted by the fire service over 20 years ago. Today the majority of communities with EMS are serviced through the fire department. In 1986 federal legislation made the fire service responsible for another major area of public safety: emergency response planning for hazardous materials incidents. In addition, the fire service has increased its fire protection activities, such as building and code inspection and public education. However, there has been little or no review of the different skills, training, expertise, scheduling, and operating procedures that are most appropriate given the new responsibilities of the service.

The incorporation of a number of new services into the daily operations and responsibilities of many fire departments requires a more modern and professional management approach. For example, accountability is not reserved for the fireground; it also relates to daily operations and the use of nonemergency resources. In addition, the complexity of managing today's fire department requires interaction with various government departments and cooperation with other governments at a local, state, and national level. A good department must be an integral part of the public service management team providing service to residents. It must be managed well and operate on par with other departments.

Another change in the fire service is the fact that some new members choose the fire service as a career whereas in the past most were "called to serve." This calling helped build a dedication to the service that went beyond that expected from a person paid to do a job. It resulted in a commitment that placed the fire department ahead of most things in a fire fighter's life. Now, however, the fire service is seen as a logical, rational career choice. Employees in the fire department are as loyal as those in other public service departments, such as public works and parks and recreation.

The fire service is now a profession, requiring professional competencies on the part of its members. Whether it is a volunteer, career, or paid on-call fire department, members are expected to receive professional training and to respond to calls in a professional manner. Fire officers must become administrators, understanding and using theories and techniques appropriate for the operation of an organization staffed by professionals.

THE STRUCTURE OF THE FIRE SERVICE

The fire service is widely accepted as a paramilitary organization. Promotion is accompanied by a change in rank, and chain-of-command authority in an emergency situation is absolute. This adherence to a paramilitary structure has focused daily operation activities on technical training and delivery of emergency service. As a result, the other aspects of an organization have not received adequate attention. Just as the military is constantly preparing for war, fire departments are constantly preparing for an emergency. However, the military has changed considerably over the past 28 years. Officers specialize in planning, communication, leadership, and purchasing procedures. Likewise, the fire service must recognize that for many departments the majority of the time is spent in nonemergency situations. Thus, the paramilitary emphasis may limit the efficiency and effectiveness of daily operations.

The paramilitary structure is already in place for most departments. It works well in emergency situations, but it must also be adapted to facilitate better nonemergency operation. For example, while on the fireground it is not possible to ask fire fighters for their opinion concerning the best deployment of apparatus. However, when setting the training schedule for the next 6 months, it is wise to ask fire fighters which training topics would be most beneficial. Likewise, during an emergency paramedics are not concerned with the relative cost of medical supplies when determining which procedures to follow. The purchasing officer, on the other hand, is concerned with relative prices from suppliers when deciding which brands to purchase.

The paramilitary structure does not need to be replaced; it needs to be enhanced. Fire departments must recognize that they perform a wide range of operational activities requiring a variety of leadership styles and administrative skills. The strict paramilitary structure must bend to allow progressive management practices to emerge.

Changes in services offered by the fire service also have brought about a change in the type of technical training and expertise necessary. The addition of EMS brought with it a requirement for greater medical training and the ability to utilize life-saving equipment. Hazardous materials response required the acquisition of new protective equipment and learning new mitigation techniques. Even public education activities required fire fighters to learn how to disseminate information to a variety of audiences, requiring skills in communication and presentation.

Increasingly, laws charging the fire service with its new services, such as Title III of the Superfund Amendment and Reauthorization Act (SARA Title III), which designates first responder training for emergency responders (including fire departments) in hazardous materials situations, dictate the type of training required. This training matches the response and mitigation responsibilities and techniques currently available. SARA Title III also designates the fire department as an information repository. Industry must report the presence of materials to the fire department. The fire department, in turn, is to use the information to prepare for a potential emergency situation.

Thus, the new services not only require new response techniques and training, but also increasingly demand more administrative and management skills. Information processing is becoming critical. Hazardous materials information is used to determine plans of attack and training topics to be covered by the department. EMS also requires new and better information management. Training and certification mandated for EMS responders must be ongoing. Certification is accompanied by continuing education requirements. This education must be documented, and the training must be scheduled according to state legislation. Testing procedures are also often mandated. Thus, it is not simply an expectation of technical training that is placed on the fire department. Management capabilities are also required in order to properly administer the new services.

A TIME FOR CHANGE

Fire administration has changed. Tradition has been modified and given way to functional pragmatism. Much of this pragmatism developed in the late 1980s and early 1990s due to reduction in available financial resources and increased demand for service. This utilitarian approach has translated into cooperative purchasing of fire apparatus and equipment, new mutual response agreements, integrated training, and the incorporation of fire districts. The reality is that in times of excess resources, or even adequate resources, surplus expenditures to maintain tradition can be allocated. In times of declining and inadequate resources, however, the cost of tradition and outmoded procedures becomes prohibitive. Public officials and city managers—and now fire chiefs—are carefully sifting through the various ways of doing business, searching for the combination of alternatives that best fits the situation at hand. Recognition has come that the combination of functions may—and in fact should—differ from one community to another. However, understanding the options, the various techniques of administration, is crucial to selecting the proper combination.

Very often public administrators, as well as members of the fire service, do not regard the fire chief as a department head. Yet the chief is in charge of important public service functions, controls a budget, directs staffing, commands resources, and provides a direct service to the public. Because of the nature of the service and the attention on emergency situations, however, the daily operations of the department are often overshadowed. As a result, administrators may not recognize the need for the same skills from the fire chief.

Increasingly, because of changes in the fire service as well as changes in public administration, there is a greater need for a fire chief to have and to *use* administrative skills. The chief can no longer rely on the budget officer to handle the department finances, for example. Hiring and firing on the basis of who you know or how someone "gets along" are no longer acceptable or legal. Fire officers must consider themselves public administrators, with comparable responsibilities and skills.

With decreases in fires and an increase in other fire department services, fire officers need to deal with personnel in nonemergency situations more frequently and more effectively. Giving direct orders to be obeyed without question may work well in an emergency, but this approach does not build mutual respect and cooperation when used on a daily basis for routine operations. Communication entails more than a loud voice that can be heard at a great distance over roaring flames. It requires sensitivity to the situation, using different vocal patterns, knowing when to listen as well as to speak, asking as well as answering questions, and formulating ideas in a manner that can be understood and accepted.

The fire service has not reached its full potential. Both technical and administrative innovations will come to the fore in the decade ahead. Fire officers must be flexible and continue to learn and develop their administrative and response capabilities. As Chief Alan V. Brunacini says, "If you like change, you will love our future."[5]

ENDNOTES

1. Much of the discussion in this section is adapted from David H. Hoover, "Preparing Fire Officers for Administrative Responsibilities," Ph.D. dissertation (Cincinnati, OH: The Union Institute, 1993), Chapters 1 and 3.
2. Ibid.
3. Richard M. Patterson, *Assessing the Veracity of Allegations of Rape: A Heuristic Model of Investigator Decision-Making* (Akron, OH: The University of Akron, 1992).
4. Hoover, "Preparing Fire Officers," pp. 37–38.
5. Alan V. Brunacini, "Chief Concerns: The Changing Role of the Fire Service," *Firehouse,* April 1993, pp. 20 and 102.

CHAPTER 2

Leadership

This chapter presents a foundation for leadership development based on an overview of proven leadership theories that are applicable to the fire service. Following the discussion of these theories, you will be able to identify your own primary leadership style using the behavioral-style inventory model. This model will help you identify the strengths and weaknesses of your leadership style and provide a basis for improving your leadership skills.

The discussion of leadership theories begins with the Leadership Grid, which uses a range of management styles to qualify leadership traits. Contingency leadership and the continuum of leadership behavior model outline leadership traits ranging from autocratic to participative styles. Follower-based theory provides leaders with six criteria to measure an individual's willingness to accept the responsibility for choosing his or her actions. Theory Z was developed to promote structuring the work environment to enhance worker productivity based on a model that includes career development and employee participation in decision making. Transformational leadership recognizes the need for departments and organizations to change in order to adapt to new challenges and societal situations. Pluralistic/participative leadership is designed to address the increasing cultural and demographic diversity of the fire service and the communities it serves.

Leadership styles set the tone for the rest of fire department operations. Leadership is a skill required at all ranks and in every fire service activity. In fire service, one can find a range of leadership, from poor to excellent. One thing is constant: a continuous need for leadership—good leadership.

Effective leadership is crucial for the successful management of a fire department whether fire fighters are volunteers[1] or paid personnel. Leadership styles are affected by the culture and tradition of the fire service. Since fire departments usually function as paramilitary organizations, there is a pronounced need for clearly defined, strong leadership roles. The application of leadership skills must expand from fireground emergencies to cover the organization, scheduling, and management of fire fighters' nonemergency responsibilities and training.

DEFINING LEADERSHIP

What is leadership? What elements combine to form the image of an effective leader? These are difficult questions to answer—there are presently over 350 definitions of leadership in management literature.[2] Leadership is most often defined as a perception or attitude held by both followers and leaders that drives the organization and causes something to happen. Some of these leadership traits include functional and organizational expertise, professional commitment, and creative problem solving, as shown in the following examples: (1) the fireground commander who gives orders at an emergency incident and joins together the necessary components required to successfully control the situation; (2) the public information officer (PIO) who establishes procedures to ensure that information concerning department operations is provided to the media; (3) the volunteer fire fighter who works full-time as a heavy equipment mechanic and chooses to use his training to maintain the department's apparatus in top running condition; (4) the fire department association secretary/treasurer who develops and directs a new fund-raising project for the department to provide adequate funding for the purchase of a new EMS unit. Although elements of leadership vary in the examples, the desire for improvement or change is a common element that distinguishes each as a leader.

Recognizing effective leaders and leadership styles can be difficult. Individual perceptions of effective or ineffective leadership may vary, depending on the specific leadership requirements for a particular situation and the culture of the individuals being led or served. Demon-

FIGURE 2-1 Leadership may be demonstrated in highly visible ways.

strating exemplary leadership does not always require a fire fighter or officer to be out in front of the company or bureau issuing orders. Leadership can be quiet and unassuming. For example, fire fighters who are quiet or reserved and don't usually offer information unless asked to do so and individuals overshadowed by the hierarchical structure of the department are often not recognized as being leaders although they may have provided impetus for the improved appearance of the fire station or identified and developed a means to improve communication between divisions in the department. Other situations require highly visible leadership for the organization to function at the appropriate service level, e.g., a hazardous materials emergency. In each situation, the effectiveness of one's leadership style can be found within the attitude of the group being led and the results their efforts have produced. Appropriate leadership style and skills should be evaluated in terms of the *leader's effectiveness and competency in influencing the activities of individuals or groups in order to achieve the goals of the organization, within the mission of the organization.*[3]

FIGURE 2-2 Leadership is also demonstrated in subtle ways, such as helping out around the station while off duty.

What defines leadership is not universally recognized. The perception of leadership is situational and often evolves from within the organizational culture. In some instances, leadership skills are identified or defined by: (1) parameters established by the organization, (2) tradition passed on from previous times, or (3) the local culture and citizens who are served by the fire department. Fire service leadership carries with it a variety of different meanings, all of which have significance in specific situations. What works in one department may need to be altered before being successful in another.

Both the definition of leadership skills and the means to acquire them are difficult to outline. Many fire service members aspire to leadership roles without having an understanding of either the skills required for such a position or the methods by which one acquires such skills. Some believe that leaders are born and only a select group of individuals can aspire to leadership positions. Others believe that leadership skills can be acquired through avenues such as formal training programs, advanced education, on-the-job training, vicarious learning experiences, and mentoring programs. Leadership skills can be acquired in any of these ways. No one system works best in all cases. Fire service personnel should explore various options and choose the method that best fits the organization.

Leadership versus Management

Fire departments are typically managed by career fire fighters (chiefs) or by appointed persons, e.g., a safety director or fire service coordinator. It is often assumed that the management decisions of these individuals define leadership because these individuals hold positions of authority. An important distinction between managers and leaders can be summarized in the following phrases: "Managers do things right" and "Leaders do the right things." Managers tend to promote efficient delivery of services; leaders emphasize effectiveness based on vision and judgment.

Leaders	*Managers*
• Answer What and Why?	• Answer How?
• Lead	• Accomplish
• Guide	• Conduct
• Influence	• Have responsibility
• Are concerned with effectiveness	• Are concerned with efficiency

This leader-versus-manager dilemma is faced by many organizations that are bound by mandatory administrative requirements and strict operating procedures. Organizational structure lends itself well to the issues of both accountability and efficiency,[4] but may restrict or inhibit discussion and examination of alternative methods of service delivery and task performance. *Management* and *leadership* are often interchanged when fire service personnel discuss activities relating to department operations. Both terms describe administrative activities and both have relevance when used to evaluate specific functions. "To manage means to bring about, to accomplish, to have charge of, or responsibility for, or conduct." "To lead is to influence or guide in the direction or course of action or opinion."[5]

Management relates to efficiency of the fire department. Leadership refers to the effectiveness of the fire department. One is pragmatic, the other is visionary. Both are needed in the fire service administration, balanced to address the goals of the organization.

Sources of Traditional Power

In order to understand the impact of leadership styles on the fire service, we must first explore the use of power. Organizations are made up of people, working together to achieve the goals or mission of the organization within established parameters. These boundaries include the number of personnel, equipment and resources, and level of service required. However, some individuals within the organization have been granted the power and authority to direct worker efforts and use available resources to achieve the stated goals of the organization. Leadership is the wise use of power if the goals are met using the available resources effectively.[6] Although power is not equivalent to leadership, it is an essential ingredient.[7]

A number of distinct types of power can affect the leadership in an organization. French and Raven[8] call them "bases of power," which include the following.

Coercive power is the leader's power to withhold rewards, punish employees, dispense undesirable assignments, and threaten. *Example: The fire administrator of a full-time fire department assigns two engine company personnel to permanent dispatch duty because they questioned the need for certain department rules and regulations. The administrator openly acknowledges that the dispatch center is the "bad boy assignment" in the fire department organizational structure.*

The leader's ability to provide both material and nonmaterial incentives is called **reward power.** *Example: The fire chief of a volunteer fire department enthusiastically encourages all personnel interested in becoming fire officers to complete the fire science courses offered by the local community college. The mayor supports the fire chief's request for tuition reimbursement and authorizes payments covering two courses per semester for each volunteer enrolled in the program.*

Legitimate power is gained through the followers' perception that the leader has the power and, therefore, it is their obligation to follow.

Example: New combination fire department part-paid recruits are required (within their first 30 days on the department) to visually identify each fire department officer by name and rank, and be able to describe their officer responsibilities. The purpose of this requirement is to ensure that all personnel become familiar with the department's administrative structure and understand each officer's responsibility and authority within the fire department.

Referent power is gained by developing sincere relationships with followers that generate feelings of admiration for the leader and a desire to associate with the leader. *Example: The full-time fire department lieutenant assigned to Station 7 is considered by the station personnel as the most caring person with whom they have ever been associated. Personnel birthdays or other special occasions are always acknowledged by the lieutenant, and there is never a request for transfer from Station 7. Personnel always trust the lieutenant's judgment in emergency situations because they admire the officer's coolness in the face of adversity. Last winter, the lieutenant used personal time to be at the hospital with a fire fighter from the station whose child was having surgery. Recently, the fire chief caused a minor uprising within the station by suggesting that company officers should rotate to different stations on an annual basis. The fire fighter/paramedics assigned to Station 7 requested that the suggestion be tabled since they respect their officer and would do anything the lieutenant asked them to do.*

Expert power is based on the leader's job-related knowledge and skills that were gained through training and education programs and experience. *Example: The combination fire department's fire prevention captain is widely respected in the community for her effective code enforcement efforts in new high-rise buildings. Recently, she was the guest speaker at a Rotary luncheon and was introduced by the fire chief as "the most powerful public official in the community," because she knew the fire and building codes better than anyone else.*[9]

Three other sources of power are information power, connection power, and persuasive power. The use of **information power** occurs when persons who have access to data or information intentionally control its flow or distribution to others within the organization. An example of this occurs when the director of public safety filters information that reaches the mayor and council. Although these persons have no official or independent power, their use of information power makes them as important as high-ranking government officials.

Connection power is closely related to information power in that it evolves from a leader's association or connection with important individuals, both inside and outside of the organization. For example, a battalion chief who is a close friend of the fire chief is more influential than other battalion chiefs who don't have the same relationship with the fire chief. The very fact that the influential battalion chief may enjoy unlimited access to the chief can in itself be construed as connection power by others within the fire department.[10]

Some individuals have the ability to convince or persuade people to do almost anything they want. They can present convincing arguments and know how to relate to different cultures and value systems. These individuals manage to accomplish things without having to give orders; they simply persuade others to do what they wish. This **persuasive power** is often held by charismatic leaders.

The leader's use of an appropriate base of power is important since it is the vehicle for motivating and influencing both peers and subordinates to achieve the goals and objectives of the department. Certain power bases may have greater appeal than others, depending on the culture of the fire department and the situation in which the power is used. For example, the use of coercive power may cause frustration and anger among followers, thereby reducing its effectiveness as a means to modify employee behavior. However, employee job satisfaction and performance are enhanced through the use of referent and expert power.

No one traditional power base or leadership style is right for every situation or every fire department. A fire officer must first understand the culture of the fire department organization and then adopt the most appropriate leadership style to meet the needs of the organization. For example, although coercive power is often viewed as negative, a fire officer may find that coercive power is a positive approach that establishes performance boundaries and expectations for a fire fighter who consistently challenges orders and ignores "requests."

Like any other organization, the fire service has a unique combination of personalities and technical expertise. Research indicates that fire fighters have unique learning styles and approaches to problem solving.[11] Leadership styles should be modified to correspond with the organizational culture as well as the learning and follower styles of each department. It is important to match the most appropriate power base to the situation because each base of power has a distinct effect when applied to fire service organizations and different cultural settings. One general problem with leadership in organizations is that individuals are not familiar with various sources of power at their command. This is important because one key to successful leadership is the ability to move from one power source to another as appropriate.

Sources of Authority

Power is the ability to get someone to do something they may not ordinarily do. **Authority** is often understood as the *right* to use power in such a manner. This inherent right is given to or earned by the individual considered to be a leader. It has different forms and sources. Authority can explain why a leadership style using the same source of power works for one fire officer and not another.

The characteristics and validity of various types of authority, such as those defined by Max Weber as traditional, charismatic, and rational–legal, have long been at the center of leadership theory discussions.[12] **Traditional authority** emanates from historically established relationships between the leader and the led. Since the fire service is a traditional paramilitary organization, this style is appropriate in emergency situations and, therefore, is prevalent in many departments. It is characterized by the members' perception that the leader's power is justified by virtue of custom and tradition.[13] Statements made by fire fighters such as "I don't like the chief personally, but I respect the position of chief" or "Our captain doesn't have much fireground experience, but we obey his orders in emergency situations" reflect manifestations of traditional authority. Respect for the position evolves from the traditional fire service paramilitary hierarchical structure. The chief, simply by being the chief, has the traditional authority or right to issue commands that are to be followed.

The **charismatic leader** gains authority from followers by force of personality and unique qualities. Charismatic leaders can often induce unquestioning loyalty and willingness to work far beyond the boundaries of customary expectations. As a result, they are often highly successful in achieving goals. However, there are two types of charismatic leaders: ethical and unethical. Ethical charismatic leaders will incorporate followers' hopes, dreams, and aspirations into their vision for the organization. They welcome feedback, share information with others, and accept criticism. This type of leadership promotes harmony among members of the organization and encourages creativity at all levels. On the other hand, the unethical charismatic leader manipulates followers, promoting what is best for his or her own self-interest and punishing creativity. (See Chapter 12, Ethics.)

The following chart is an easy reference for identifying ethical and unethical charismatic leadership styles.

Unethical charismatic leader	*Ethical charismatic leader*
• Uses power only for personal gain or impact	• Uses power to serve others
• Promotes own personal vision	• Aligns vision with followers' needs and aspirations
• Censures critical or opposing views	• Considers and learns from criticism
• Demands own decisions be accepted without question	• Stimulates followers to think independently and to question the leader's view
• One-way communication	• Open, two-way communication
• Insensitive to followers' needs	• Coaches, develops, and supports followers; shares recognition with others
• Relies on convenient external moral standards to satisfy self-interests	• Relies on internal moral standards to satisfy organizational and societal interests

The third type of authority is **rational–legal,** which derives its power from the follower's perception that authority is based on rules and regulations considered either legal or correct. This style has strong roots within the fire service since much of the command and control structure is based on: (1) department rules and regulations, (2) state or federal laws, (3) standard operating procedures or guidelines, and (4) labor contracts. It enforces the notion that followers should adhere to the fire department pronouncements or face the prospect of being punished.[14]

A variation of the Weber leadership theory has been set forth by behaviorist Kurt Lewin.[15] This theory identifies three distinct kinds of leadership: autocratic, democratic, and laissez-faire. **Autocratic** leadership is characterized by tight control of organization activities where members are not permitted to participate in decision making. The autocratic leader centralizes power and issues absolute orders. Use of the autocratic leadership style in the fire service is most appropriately employed on the fireground or in other emergency situations where orders must be followed implicitly. However, it may not be appropriate when used to guide the fire department's other daily operations and nonemergency activities.

The opportunity for members to provide input into decision making is a shared element of pluralistic, charismatic, and democratic leadership styles. **Democratic** leadership encourages follower (group) participation in making decisions and promotes open discussion of fire department member concerns and ideas. Some management decisions are made by majority rule. The democratic form of leadership may seem to be the most cohesive style; however, it may be inappropriate for emergency operations. For example, should the fireground commander (FGC) at a major structure fire promote member participation when deciding to change from an offensive to a defensive mode? The FGC uses input from other officers on the scene, but the ultimate decision concerning the mode of attack remains with the fireground commander.

Laissez-faire leaders allow the group members to do pretty much as they please. Group members are responsible for making all of the decisions based on current conditions. The laissez-faire leader will provide input and information but will not attempt to direct or guide

FIGURE 2-3 Democratic leadership, which utilizes the contributions of many members of the department, can be effective in nonemergency situations.

the group. What needs to be done may be identified, but the specific means to accomplish the tasks is left to the members. In some instances even prioritizing required tasks is left to the members of the organization. At first glance, this style of leadership does not appear to have any relevance to the fire service, but it may apply when training subordinates in decision-making procedures. Certainly, it has no place on the fireground since its implementation may spawn "freelance fire fighting," which will often work at cross purposes with the other operating units at the emergency.[16]

LEADERSHIP STYLE AND CHARACTERISTICS

Leadership style describes the various characteristics of a specific type of leader. Numerous studies have been conducted to determine what (if any) leadership style is best for all conditions. For example, the Ghiselli study, conducted by Professor Edwin Ghiselli, examined over 300 managers to identify the most prominent traits of business managers.[17] The findings concluded that six traits were significant for effective leadership:

1. Supervisory ability: getting the job done through others
2. Occupational achievement: being motivated to work hard to achieve
3. Intelligence: reasoning, judgment, and the ability to think
4. Decisiveness: being able to solve problems and make decisions
5. Self-assurance: having confidence in one's own ability
6. Initiative: being self-starting and self-directed

Successful leaders cannot be identified by isolating certain physical and personality traits. However, certain traits are desirable in most leadership situations.[18] Although the Ghiselli study is significant, different research has shown that other traits are equally important, including:

- Persistence
- Mental energy
- Integrity
- Persuasiveness
- Ability to handle people
- Self-confidence
- Mental and physical endurance
- Enthusiasm
- Sense of responsibility
- Desire to achieve[19]

One must become familiar with recognized leadership theories and styles to be able to identify the most appropriate leadership style to apply in a particular situation.

Leadership theory provides an explanation of what does and does not work and why.[20] Styles and theories can be differentiated from one another, but there is no one best leadership style or theory that applies universally. As an example, Rue and Byars believe that work-centered leaders practice an autocratic leadership style, while employee-centered leaders tend to practice a more democratic style of leadership. They conclude that most leaders are not all work-centered or employee-centered, but occupy a middle ground that is a mix of the two extremes.[21] The mixture of styles is dependent on issues such as personality, training, experience, and situation. These may be altered to suit the circumstances or the personnel involved. Each style must be evaluated in the context of each situation to arrive at the best mix of leadership styles to accomplish the goals of the organization. It must be remembered: no one leadership style is best for all situations.

Overview of Pertinent Leadership Theories

Many theories have been developed to explain why leaders behave as they do, as well as the activities or characteristics of a leader. The following discussions of common leadership styles are accompanied by a brief description of how they have been observed in the fire service. These styles were selected for discussion because of their prevalence in contemporary leadership literature or their relevance to the fire service. Some of those chosen are classic theories and been proven accurate repeatedly. The reason for including each style will be explained at the beginning of the discussion. As you read each, try to think of an occasion when you have seen the style demonstrated in your community or department.

The Leadership Grid ®

The Leadership Grid, a registered trademark of Blake and Mouton, was originally created in 1964 as the Managerial Grid. It was republished in 1991 as the Leadership Grid and is the subject of a new leadership book.[22] It is included in almost every management and administration text that deals with contemporary leadership development.[23] The grid is also widely used in leadership and administrative development seminars and projects in the public sector.

The Leadership Grid is based on two leadership qualifiers called **concern for production** and **concern for people.**[24] It uses a two-dimensional grid to identify and relate different styles of leadership. Managers and supervisors are asked to complete the grid form designed to indicate how they would handle specific situations. The Leadership Grid model is intended to guide supervisors in identifying their management leadership style and developing a plan of action for moving toward a team management style.

The five major styles identified in the model include:

1. *The impoverished manager*—has low concern for both production and people. *Example: The fire officer who just doesn't*

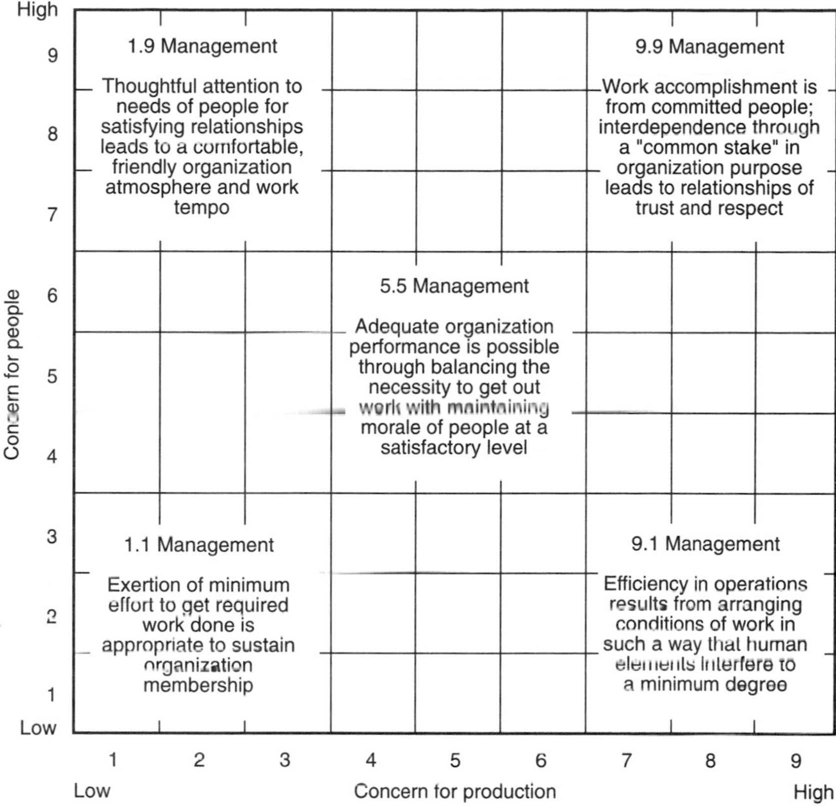

FIGURE 2-4 The Leadership Grid®

care about anyone or anything, and never participates enthusiastically in company training or other department activities. The officer is always quick to criticize others and finds fault with everything that is passed down through the chain of command.

2. *The sweatshop manager*—has high concern for production and low concern for people. *Example: The assistant chief who orders department personnel to initiate an offensive attack in a commercial structure fire even though the fire department does not have either the personnel or equipment available or on-scene to accomplish the task.*
3. *The country club manager*—has high concern for people and low concern for production. *Example: The fire officer who is usually more concerned with the well-being of department per-*

sonnel than the actual duties they are called upon to perform. Subordinates usually respect the officer's position; however, they may also learn to manipulate the country club manager to avoid station chores or other duties since the officer is not concerned with production.

4. *The organized manager*—has a balanced concern for people and production. *Example: This fire officer is capable of evaluating assigned tasks and department personnel to arrive at the most advantageous solution based on department resources and level of need.*
5. *The team manager*—has high concern for both production and people. *Example: This fire officer is always involved with department personnel both at the station and on the fireground. The power base for this officer is a blend of both* ***referent*** *and* ***expert*** *powers. "High energy" usually describes this person.*

The results of the grid scoring indicate where a person is on the Leadership Grid. In most instances, the majority of the supervisors participating score somewhere in the middle rather than at either extreme of the grid. For Blake and Mouton, the ultimate goal is to help supervisors achieve a high concern for people and production. Thus, a fire officer is concerned both with the quality and maintenance of the apparatus and with the well-being and morale of the fire fighters in the department. Blake and Mouton describe this as a team approach to management that includes various members of the department in management decisions and operations.

Contingency Leadership

The Leadership Grid is an attempt to identify one best leadership style that fits all situations and individuals. However, not all successful leaders fit the definition of "enlightened managers" as defined by the Leadership Grid. We all know of situations where task-oriented (production) leaders were successful. It is generally recognized that under emergency or high-pressure situations, task-oriented leadership is most often preferred by subordinates. On the other hand, subordinate- or employee-centered leaders can be equally successful. The Leadership Grid provides a way to identify and classify different styles that fire officers may encounter.

The contingency leadership theory was developed in the late 1960s when theorists wrote that there was no one best leadership style for all situations.[25, 26] Also called situational leadership, this theory maintains that the appropriate leadership style varies from one situation to another and with differences in follower characteristics. Contingency leadership utilizes various environmental, technological, and organizational characteristics of successful leadership styles.[27]

The fire service has begun to recognize the value of the contingency leadership style. For example, the fire prevention and education efforts of the fire service have contributed to the decline in the number of fires in the U.S. in the last decade. Supporting legislation to require the installation of residential smoke detectors and residential sprinkler systems in single- and multi-family dwellings[28] required persuasive leadership working with businesses in the community. Fire-department-sponsored public education programs require persuasive and participatory leadership to improve community awareness and behavior. In addition, the reduction of financial resources has forced many fire departments to downsize their staffing requirements and reassess the level of service provided to their communities. This process requires negotiation with the community and fire department personnel. Successfully negotiating these changes requires the fire officer to employ a more democratic leadership style.

Improved fire protection and fire safety education and reduced resources are three examples of change that have compelled fire administrators to search for the best method of leading an organization in a dynamic environment. The most appropriate leadership style to use will depend on the specific circumstances surrounding the department. Contingency leadership has been embraced by many fire administrators as the most viable leadership model for the changing times. The successful fire department of the future will be led by flexible administrators who adjust their leadership styles to correspond with the situation at hand and the personnel involved. This is challenging, as it requires leaders to change their personal styles as necessary in order to be most effective in different situations.

Continuum of Leadership Behavior

The continuum of leadership behavior model was developed by Robert Tannenbaum and Warren Schmidt (see Figure 2-5).[29] Based on the contingency leadership theory, the continuum expresses a range of actions a leader might take, from most autocratic to most participative, as appropriate to the situation. For example, the autocratic fire officer who makes a decision to purchase new personal protective gloves without consulting other officers or fire fighters demonstrates a "boss-centered" leadership style.[30] The leader using the participative style will permit subordinates to provide input into the decision within the limits established by the fire officer. In this case, the fire fighters would be asked to share their experiences with the gloves currently in use and recommend features they would like to see in the new gloves. In some instances, the fire fighters would be asked to field-test various brands and report the results to their officers. Thus, the fire officer analyzes the situation in terms of status of the above criteria and selects the leadership style accordingly.

Successfully using the continuum of leadership model requires the fire officer to carefully select the most appropriate leadership style based on the current situation and the degree of leadership or management control desired. This is determined in part by the amount of discretion allocated to the fire officer and what he or she chooses to share with subordinates. Time may also influence selection of style, as will accountability.

Follower-Based Theory

One of the most popular leadership theories among administrators is the follower-based theory formulated by Hersey and Blanchard. Although it shares many similarities with other contingency or situational leadership theories, it identifies task maturity as the most important variable. Task maturity is defined as "the ability and willingness of the people to take responsibility for directing their own behavior."[31]

FIGURE 2-5 The Leadership Continuum

Autocratic Style

1. Leader makes decision and announces it
2. Leader "sells" decision
3. Leader presents ideas and invites questions
4. Leader presents tentative decision subject to change
5. Leader presents problem, gets suggestions, makes decision
6. Leader defines limits and asks group to make decision
7. Leader permits subordinates to function within limits defined by leader

Participative Style

a. Scale ranges from totally autocratic to totally participative

b. Most appropriate category is determined by three factors:
   - The supervisor's preferred style
   - The subordinates' preferred leadership style–what makes them most comfortable
   - The situation at hand

c. Third factor is most influential since it can impact the subordinates' preference

Task maturity is dependent upon: (1) past job experience, (2) relevant job knowledge, (3) understanding of job requirements, (4) a willingness to take responsibility, (5) organizational commitment, and (6) achievement motivation. Personnel who possess high levels of these attributes are deemed mature, while those who lack such traits are immature. (The terms maturity and immaturity are merely classification techniques.) Therefore, managers should alter their leadership style to reflect the task maturity level of the employees they are supervising.

The follower-based theory is relevant to fire service leadership since personnel assignments (either company or bureaus) are often based on a seniority system or group selection process that does not consider task maturity elements. (Seniority may positively influence past job experience and relevant job knowledge but has no bearing on items 3 through 6.) It is important that an officer evaluate all personnel based on the criteria identified in this model. By doing so, the fire officer is better able to understand the need to adjust leadership styles to match the situation and the various characteristics of fire fighters and junior officers.

Theory Z

In recent years, the business community has come to associate the letter "Z" with leadership. Theory Z, derived by William Ouchi in 1981, is the leadership model that best describes those management practices that brought success to Japan's business and industrial communities for many years. The basic premise of Theory Z is that workers are the key to high productivity and, therefore, management's role is to structure the work environment so that employees can work together more effectively. Ouchi argues that many of the Theory Z work attributes are missing from American organizations. Based on his research at several major U.S. companies, Ouchi integrated successful business practices in America and Japan into one middle-ground model appropriate for use in the U.S. The seven important dimensions of Theory Z are:

1. Long-term employment (job security)
2. Collective decision making

3. Individual responsibility
4. Slow evaluation and promotion
5. Implicit, informal control with explicit, formalized measures
6. Moderately specialized career paths
7. Holistic concern, including family[32]

It is interesting to note that many of the employee-based quality aspects of Theory Z are based on the works of John Deming, an American theorist who could not convince American managers to try his ideas. Proven effective in Japan, however, Deming is finding a very interested audience in the U.S.[33]

Some critics have suggested that Theory Z emphasizes values and attitudes endemic to Japan, and it is therefore unreasonable to expect general acceptance in the U.S.[34] Because of the sociocultural, economic, and technological differences between the two countries, it is not feasible to implement the Japanese management style in the U.S. However, many organizations, such as IBM, Eastman Kodak, and Hewlett-Packard, have integrated some of William Ouchi's suggestions into traditional American leadership models and formed a modified system that satisfies their specific organizations.[35]

Several Theory Z dimensions are evident in the fire service and the way it operates. Long-term employment can relate to both career and volunteer departments since people associated with these departments often dedicate many years to the vocation or avocation. Slow evaluation and promotion is an important element in the fire service since promotions are usually based on seniority, performance evaluations, or a combination of knowledge and experience. In addition, there are relatively few officer positions available in fire departments. The structure of the fire service is a relatively flat pyramid, with few levels of promotion offered. Career advancement tends to be deliberate and cautious and is part of the tradition of the American fire service.

Moderately specialized career paths are the mainstay of the fire service. Many volunteer and career personnel prefer one type of job function or activity over another. This phenomenon tends to organize the fire department into small areas of specialization, whether they are formal or informal. For example, some members will specialize in auto extrication and rescue problems while others dedicate their effort

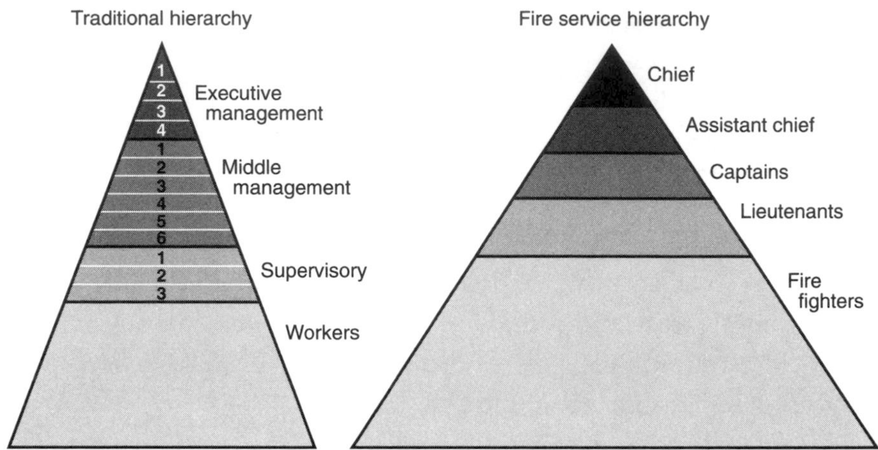

FIGURE 2-6 The hierarchy of the fire service tends to be flatter than that of other organizations, and there are fewer positions in each subsequent level.

toward effective rural water shuttle systems. Each becomes an area of specialization that can translate into career paths.

Finally, holistic concern is an integral part of belonging to the fire service. Holistic concern goes beyond the fire department and job assignments. It includes: (1) caring about each other's families, (2) being concerned about the community at large, (3) having genuine concern for other members of the department, and (4) maintaining the family atmosphere that is cultivated by members who share specific shifts or who work closely with other members for long periods of time. This holistic commitment has been a central part of the volunteer fire service in America, where the fire department is also a center for family and community activities. The holistic atmosphere is sustained in many combination fire departments. Shift assignments in full-time fire departments promote the feeling of family and extend concern for one another away from the fire department.

While not all elements of Theory Z are appropriate for emergency service agencies, the collective decision-making element may prove to be a means of increasing employee participation in the management process. Fire administrators should view the Theory Z leadership model as another tool to better lead and manage the department.

Transformational Leadership

Transformational leadership is of particular relevance to the fire service because the fire service as a whole is in transition. Duties and responsibilities have expanded, and the types of services provided are more highly specialized. In addition, financial constraints require innovative staffing patterns and resource allocation. Transformational leadership can help guide fire departments through these changes.

Transformational leadership is about change, innovation, and entrepreneurship. The focus of this model is accomplishing positive change–the transformation of organizations. While financial acumen, manufacturing expertise, and marketing prowess are important ingredients in the success of most organizations, they are not sufficient for organizational transformation.[36] Transformational leaders challenge the organization, perform, and move it through three acts:

Act I. *Recognizing the Need for Revitalization.* The leader recognizes the need for change in order to keep up with the rapidly changing environment.

Act II. *Creating a New Vision.* The leader envisions a changed organization and motivates people to embrace the vision and make it a reality.

Act III. *Institutionalizing Change.* The leader guides people in making the vision a reality.

Transformational leaders are usually courageous individuals who: (1) believe in people, (2) are willing to take risks, (3) are value driven, (4) can deal with complexity, ambiguity, and uncertainty, and (5) are lifelong learners. They are visionaries![37]

In Search of Excellence and *Thriving on Chaos* are two books by Tom Peters that point out the importance of an organization's ability to change. Successful managers discussed in these books were visionaries who could motivate their followers to see and accept their vision of the future. Peters identified, among other things, the need to take risks and try new things, and also the ability to work in an unstructured and possibly unstable environment, as keys to success.[38] These are precisely the elements identified in the transformational leadership theory.

Pluralistic Leadership

The benefits of and need for organizational transformation, coupled with the realization of the changing nature of the workforce, as documented in *Workforce 2000,*[39] led to the development of the pluralistic leadership theory. Both pluralistic leadership and participative leadership emphasize empowerment and employee involvement. A pluralistic leader recognizes the need for a change in organizational culture to create an atmosphere that values cultural diversity. As such, a pluralistic leader incorporates aspects of transformational leadership. This change in culture must be a collaborative process involving all members of the department, both minority and majority members. The fire officer can inspire commitment to change on the part of the members of the organization by setting an example and institutionalizing change.

Marilyn Loden and Dr. Thomas A. Gordon developed a model of pluralistic leadership based on research in twenty public and private organizations.[40] The primary characteristics of this pluralistic leadership model are as follows:

1. Vision and values that recognize and support diversity within the organization
2. Ethical commitment to fairness and the elimination of all types of workplace discrimination
3. Broad knowledge and awareness of the various and subtle dimensions of diversity and multicultural issues
4. Openness to change based on diverse input and feedback about personal filters and blind spots
5. Mentoring and empowering diverse employees
6. Being an ongoing catalyst and model for individual and organizational change

Pluralistic leadership expects the visions of the leader to incorporate the reality of the changing nature of society and the workforce. Many communities served by the fire service are very different from those served 15 or even 10 years ago. Widespread immigration has changed the cultural atmosphere of cities, suburbs, and small towns across the country. (See Chapter 3, Cultural Differences, for further discussion.) Pluralistic fire officers will value the differences brought to the organization by diverse employees. This will translate into an ethical commitment to fairness that goes beyond meeting the letter of the laws prohibiting discrimination. The pluralistic leader strives to do what is *right* as opposed to taking only those steps that are *legally required.*

Leaders need to understand their subordinates. Thus, a pluralistic leader must have a broad understanding of diversity and multicultural issues. Taking a proactive approach to broader understanding is crucial to pluralistic leaders. They must take the initiative to gain the requisite knowledge. This initiative demonstrates the importance of the effort to other supervisors, leaders, and officers in the fire service. Understanding leads to tolerance, which leads to recognizing contributions, which leads to change. This is a great challenge to the fire service because much of what it does is steeped in tradition. However, successful organizations of the future will learn to do the same thing in different ways. They will learn to change how services are delivered. This requires willingness to learn from minority groups with ideas that are not shaped by the cultural traditions of the current majority of fire fighters.

As part of the clear demonstration of commitment to a diversified organization, a pluralistic leader serves as a mentor to members of various cultural groups in the organization. (See the section on Mentoring in Chapter 6.) In a mentoring role, the fire officer needs to learn as well as to teach in order to ensure that the contributions from the protégé's culture are valued and maintained. Employees from diverse cultures are empowered to make contributions, to share ideas, and to present alternative ways of doing business. Rather than smothering or ridiculing these differences, the pluralistic fire officer encourages innovation and experimentation. The leader must demonstrate personal commitment to integrating new ideas and serve as the catalyst for others in all levels of the organization. This respect for difference must permeate the fire department in order to be truly successful. Fire fighters and paramedics must learn to value the ideas and concepts brought to the department by members of different cultural groups, as well as by women. This leadership style is one that looks forward and considers the reality of the changing workforce and the changing community the fire department is serving.

Identifying Your Primary Leadership Style

The theories presented in the preceding section provide a background for understanding leadership. Using this foundation, fire officers can begin to assess and improve their personal leadership capabilities. In order to facilitate this, the Behavioral Style Inventory Model is presented. This model is a tool to assist in applying leadership theories to daily fire department operations. It is not a substitute for understanding the various theories.

Being a leader is more than simply giving orders at an emergency scene or making operational decisions that affect the future direction of the department. It encompasses our attitude in relating to other members of the department and to the general public. In everyday communication and interaction with others, each individual tends to utilize a specific approach that is most comfortable to that individual. This is referred to as a behavioral style.

Behavioral styles are general descriptions of the methods of interaction that individuals utilize most frequently. Everyone uses different styles at various points in time, but one style tends to dominate.

Being aware of behavioral styles can enhance your ability to interact with various individuals. You can begin to recognize why you work well with some individuals and not as well with others.

According to Dr. N. Ted Oravitz of Performance Development Associates of San Antonio, Texas, there are four behavior styles that impact daily interaction: **director, persuader, supporter,** and **analyzer.**[41] In addition, there are four interpersonal communications characteristics that help describe your style. The degree to which each of these characteristics is part of your daily interactive pattern determines your behavioral style. As you read them, ask yourself which ones best describe your style.

1. *Decisive.* Having the power or quality of deciding. Marked by or indicative of determination or firmness. Resolute, unmistakable, unquestionable.
2. *Cautious.* To be on guard. Prudent forethought to minimize risk. Prudently watchful and wary.
3. *Task.* Orientation focuses on the assigned piece of work; something that has to be done. Assignment, task, duty, job, chore, stint.
4. *Relations.* To have relationship or connection. To have or establish a relationship. To respond favorably. The state of being mutually or reciprocally interested.

In everyday activities, we observe that some fire fighters are highly assertive, while others are more responsive. There are those who pride themselves on being task oriented and still others who tend to act independently and don't concern themselves with relationships.

What is the best method of coping with different behavior styles? You must ask yourself the question, How do I adapt my leadership style to each situation? The leadership models previously discussed in this chapter identify a variety of styles and methods that have been successfully implemented by managers. But what is best for your situation? In order to make that determination, you must first examine your leadership approach to determine your **dominant behavior style.** For example, do you enjoy analyzing fire and EMS run data more than making quick decisions? Would you rather make decisions slowly because of your concern for others' interests, or would you rather work on a department project that is so interesting that you are caught up in the excitement of it and extend your enthusiasm to other members of the department? By identifying your dominant and secondary behaviors, you have taken the first step toward adapting yourself to different situations. In some instances these behavior styles may be congruent; in others they may be in conflict.

In order to better understand how to identify **leadership style,** the following descriptions have been provided by Dr. Oravitz. As you read each, take notes as to which profile(s) best describe you and which have no relationship to your style.

The Director (Profile)

In a typical work environment, Directors tend to operate in a quick, decisive, and businesslike manner. They manage their time efficiently and usually concentrate on the task with the highest priority rating.

Directors are efficient and effective decision makers and are prepared to take the risks necessary to move a project to completion within the parameters established for it.

Directors are very comfortable making decisions independently and can be vigorous in defending their decisions.

Directors frequently are impatient with others who are slower to make decisions. They can be brusque, blunt, and even abrasive. Most often, they have little time for "small talk" or general conversation when a specific task is before them. The task is primary; relationship with others is secondary.

In their particular work environment, Directors are aggressive, competitive, and totally committed to their goals.

The Persuader (Profile)

In their typical work environment, Persuaders tend to operate in an energetic, enthusiastic manner. Compared with Directors, Persuaders generally manage their time less efficiently and usually concentrate on the tasks they find most interesting.

Because they are habitual talkers who generate contagious enthusiasm, they get others caught up in the excitement of the project and do get results.

Persuaders are expressive extroverts who want people around them primarily as an audience. They will listen to the suggestions and ideas of other people only to the extent that they reinforce the thoughts or decisions that they (the Persuaders) already have made.

Persuaders spend excessive time in conversation, and because they generally are rather poor listeners, they do most of the talking.

In their work, Persuaders compete enthusiastically and emotionally, and they almost constantly sell themselves to the rest of the world.

The Supporter (Profile)

On the job, Supporters generally seem to operate in an empathetic and supportive manner. Compared with Directors, they manage their time less efficiently and usually concentrate on the task that is of most interest to the majority of their particular group.

Decisions are made much more slowly by Supporters, primarily because they are concerned about each individual's interests and feelings. Supporters are more concerned with comfortable relationships than tasks and are usually looking for consensus.

Supporters are considerate and empathetic people who, in order to be most effective, need a warm and friendly working climate. They are good listeners and make an effort to encourage ideas and input from others.

Supporters will spend a great deal of time and effort in building good relations with others because they feel the need to belong and be recognized.

In the work environment, Supporters compete, but somewhat hesitantly because they try to avoid stepping on other people's toes.

The Analyzer (Profile)

In the typical work environment, Analyzers tend to operate in a methodical and systematic manner. While they don't necessarily come from the same mold as the absent-minded college professor, they usually do not make the best use of their time. They tend to concentrate and spend most of their efforts on the tasks that require comprehensive analysis and clarification.

Analyzers, much of the time, are slow to make decisions. As they strive for perfection, they are constantly on the lookout for alternatives. They will submit each new idea or step to rigid testing and try to obtain documented proof.

Analyzers are people who require factual data laid out in an orderly and detailed manner, and because they always seem to be looking for additional information they tend to be indecisive.

Where the task requires a systematic step-by-step approach, Analyzers are most effective. However, time frames must be established to achieve results.

In the actual work situation, Analyzers compete in a predetermined manner while requiring clarification and confirmation at each step of the process.

To understand the impact of each of these behavior styles, we must also recognize their strengths and weaknesses. These specific strengths and weaknesses will assist you in identifying your overall style, as well as that of your superiors, and offer clues as to the type and method of behavior that can be expected from each.

Director

Strengths	*Weaknesses*
This individual usually:	**This individual may not:**
• Gets results	• Analyze pros and cons
• Stimulates action	• Weigh risks
• Accepts challenge	• Use caution
• Makes decisions	• Consider before deciding
• Acts with authority	• Protect others
• Stirs things up	
• Solves problems	
• Organizes well	

Persuader

Strengths	*Weaknesses*
This individual usually:	**This individual may:**
• Makes a good impression	• Vacillate occasionally
• Talks forcefully	• Talk too much
• Shows enthusiasm	• Not listen
• Talks easily	• Overlook facts
• Entertains	• Overpower others
• Motivates others	• Turn on and off quickly

Supporter

Strengths	*Weaknesses*
This individual tends to be: • Loyal • Cooperative • Sensitive • Consistent • Honest • Emotional • Good listener • Patient	**This individual may:** • Be thin skinned • Be easily hurt • Not talk much • Be overly sensitive • Try too hard to please • Not be assertive

Analyzer

Strengths	*Weaknesses*
This individual usually: • Follows directions • Likes organization • Thinks clearly • Acts methodically • Likes detail • Wants accuracy • Complies with authority • Shows caution	**This individual may:** • Prolong decisions • Be too conservative • Not delegate • Be too cautious • Be buried in detail • Dislike unpopular decisions

Now that you have identified your style(s), identify those of your supervisors. What types of leaders are they? Do the same with your subordinates. The purpose of this presentation is to help you learn how to get along better with others—superiors, subordinates, and other officers of equal rank. Now that you have correctly identified the styles of the members of the fire service with whom you interact on a daily basis, what should you do?

The next step is to review the Do and Don't action lists that will assist you when relating to any of the four profiles. These simple steps provide a basic system that can improve your ability to work with others in the departments, to achieve goals, and to accomplish tasks.

When Relating to a Director

DO	*DON'T*
Be clear and to the point	Be overly friendly
Avoid rambling	Generalize
Stick to the subject	Talk too much
Be logical in presenting facts	Make insupportable statements
Provide options	Get emotional
Ask pertinent questions	Be loud or boisterous
Remember that the personal relationship is less important than the task	
Establish a time frame	

When Relating to a Persuader

DO	*DON'T*
Be open, warm, and friendly	Be restrictive of their time
Concentrate on people aspect	Show "cold" manner
Take time, socialize	Do all the talking
Keep on track—diplomatically	Jump to facts too quickly
Encourage and support enthusiasm	Restrict suggestions
Be sure decision is made	
Be attentive listener	

When Relating to a Supporter

DO	*DON'T*
Be sincere	Be overpowering
Give warm, personal comments	Demand or dominate
Ask questions	Push ideas too aggressively
Listen attentively	State too many facts
Make the climate comfortable	Ask closed-end questions
Find common ground	Make communications one way
Give assurance	
Keep communications low-key	

When Relating to an Analyzer

DO	*DON'T*
Take your time	Generalize about details
Have all the facts	Be vague
Stick to the facts	Be casual
Be organized	Waste time on casual conversation
Be formal in presentation	Jump to bottom line too quickly
Be thorough	Jump around from one point to another
Avoid gimmicks	
Concentrate on specifics	
Establish a time frame	

Understanding the style of your superiors and subordinates can assist you in working with them. Use the list of do's and don'ts to develop an approach that will make interaction smoother and more effective.

Adaptability is the most crucial element in implementing and using the behavioral style inventory model. Fire officers should develop the ability to adjust to others' needs in order to achieve mutual understanding, commitment, and satisfaction. They must understand, accept, and reconcile personal issues to the opinions and viewpoints of others, regardless of cultural differences or socioeconomic status. Successful fire administrators should invite input and open discussion for the purpose of creating mutual trust and understanding. The result will be improvement at all levels of operation. Finally, officers must strive to improve personal and interpersonal relations while maintaining and promoting cooperation and accountability through the department.

SUMMARY

The behavioral style inventory model is a tool to assist fire service leaders to work better with other individuals. It goes beyond the contingency and situational theories in terms of specificity and adaptability. It does, however, also lead to the general application of other leadership theories.

It is impossible to analyze everyone and every event. When dealing with everyday situations, there tends to be a general approach that works best for each individual in the department. This is determined to some extent by the power and authority possessed by the individual fire officer. A junior fire officer with little traditional or legitimate authority based in his or her position must rely more on expert and persuasive powers to get things done. The proper use and balance of power and authority, coupled with the understanding of various theories and how to apply them, contribute to the development of effective leadership.

The information in this chapter can be used to assist fire service officers in developing their leadership style and capabilities to effectively guide their department into the future. Leadership needs continuous refreshing and updating. For example, good fireground operations command warranted promotion to chief 10 or 15 years ago. Today, the demands of dealing with financial challenges, legal requirements, and new fire fighters from various cultural backgrounds require a different type of leader. Fire officers must adapt and change. As the fire service develops, officers, too, must develop their skills as leaders and serve as guides to progress.

The fire service is in a state of constant transition and depends on sound, well-trained leaders to plan and guide its future direction. The most critical leadership element of the fire service today is the ability to respond to changes in the status quo of American culture. While other municipal agencies are in the process of assessing the impact of change on their departments, fire service leaders have an opportunity to do the same. Although tradition is an important part of the fire service, tradition combined with adaptability and responsibility reflect the future needs of effective leadership.

Fire service administrators and officers of the future must be able to adapt their leadership style to the situation at hand, without sacrificing the traditional fireground command leadership roles necessary in emergency conditions. In order to achieve this change successfully, present and future leaders must now begin to transcend the traditional "emergency mode" approach to one of understanding, accountability, responsibility and adaptability. Understanding and applying leadership theories and styles provides a framework to work better with others, both superiors and subordinates, and to make a difference in the evolution of the fire service.

CASE STUDIES

Case #1

Fire Chief Jeffrey of the Volunteer Fire Department recently appointed Paulette Miller as the EMS captain to replace the previous captain who accepted a job in another state. Although Captain Miller has only been a member of the department for 5 years, she is considered by many to be the most proficient EMS person on the department. She is somewhat quiet, but enjoys praise and recognition of her EMS performance. Because she has never served in a leadership position, Miller is concerned that others will not take her seriously as a leader. Captain Miller requests a meeting with Chief Jeffrey to discuss ways in which she can assume the new leadership role without alienating other EMS personnel and yet demonstrate that she is in charge and does have a power base.

1. What types of power does Captain Miller currently demonstrate?
2. Which base(s) of power should Captain Miller utilize in her new position? (There are more than one.)
3. How could Chief Jeffrey assist Captain Miller in building her power base? What sources of power would be wrong?
4. What leadership style would benefit Miller most?
5. What leadership style would allow Jeffrey to assist Miller?

Case #2

Assistant Chief Roberts attends a hazardous materials conference (at department expense) in order to gain new information and knowledge pertaining to community emergency response planning. Roberts is responsible for the department's SARA Title III program and also serves as the fire service representative on the Local Emergency Planning Committee (LEPC). Although he has been with the department for 16 years, and assistant chief for 5 years, he often feels threatened by the perceived aggressiveness of younger members of the department.

While at the conference, Assistant Chief Roberts attends a number of workshops that address issues pertaining to hazardous materials planning and community preparedness. He is able to speak with a number of well-informed persons in the field and obtains copious amounts of literature and other information that will enhance the department's planning efforts and assist the LEPC Facility Planning Committee. In addition, he joins a networking group comprised of other fire officers with similar hazardous materials planning interests. Upon his return to the department, several fire fighters on his shift ask about the conference and request to see the information he has gathered. Several express an interest in becoming involved in the department's hazardous materials planning process and learning more about the LEPC. The fire fighters' requests are perceived by Roberts as a threat to his leadership and power since "he is in charge of hazardous materials planning" and only he can participate in department and community activities related to hazardous materials.

After several requests from the fire fighters, Roberts agrees to provide the information; however, he selectively chooses the information for distribution, making available only basic details. Much of the information he provides has already been acquired by the fire fighters through training programs or is readily available in monthly fire service publications.

1. Which power base(s) were used?
2. What is the strength of each power base?
3. What are the weaknesses? (Hint: The use of the power base is the problem.)
4. Do the fire fighters have a power base?
5. Are they challenging the assistant chief?
6. What sources of authority validate the assistant chief's use of power?
7. What leadership style is demonstrated?

ENDNOTES

1. Justin J. Thompson, "Recruitment and Retention of Volunteer Fire Fighters," presented at the Annual Meeting of the National Fire Protection Association, New Orleans, LA, May 18, 1992, p. 29.
2. For examples of various definitions of literature, see: Gary A. Yukl, *Leadership in Organizations* (New Jersey: Prentice Hall Inc., 1981), pp. 67–91; Harold Gortner, J. Mahler, and J. Nicholson, *Organizational Theory: A Public Perspective* (Chicago: The Dorsey Press, 1987), pp. 290–300; George J. Gordon, *Public Administration in America* (New York: St. Martin's Press, 1982, 2nd edition), pp. 269–284; Burt Nanus, *Visionary Leadership* (San Francisco: Jossey-Bass Publishers, 1992); and Richard Lynch, *Lead! How Public and Nonprofit Managers Can Bring Out the Best in Themselves and Their Organizations* (San Francisco, Jossey-Bass Publishers, 1992).
3. Nicholas Henry, *Doing Public Administration: Exercises in Public Administration* (Dubuque, IA: William C. Brown, 1991), p. 65, footnote.
4. Cole Blease Graham, Jr., and Steven W. Hays, *Managing the Public Organization* (Washington, DC: Congressional Quarterly, Inc., 1986).
5. Warren Bennis and Burt Nanus, *Leaders: The Strategies for Taking Charge* (New York: Harper & Row, 1985), p. 21.
6. Bennis and Nanus, *Leaders,* p. 17.
7. Robert N. Lussier, *Supervision: A Skill Building Approach* (Homewood, IL: Irwin, 1989), p. 374. The Ghiselli study is part of an overview of leadership traits utilizing supervision.
8. John French and Bertram Raven, "The Basis of Social Power," in *Organizational Behavior and Management,* Henry Tose and W. C. Hammer, ed. (Chicago: St. Clair, 1977), pp. 442–456.
9. Lussier, *Supervision,* p. 375.
10. Graham, *Managing the Public Organization,* p. 156.
11. Peggy Linden, course development director at the National Fire Academy, is utilizing the Myers-Briggs Personality Profile instrument to examine personality and learning traits of members of the fire service.
12. Max Weber, *The Theory of Social and Economic Organization* (Glenco, IL: Free Press, 1947). For further discussion of this concept, see George J. Gordon, *Public Administration in America* (New York: St. Martin's Press, 1982, 2nd edition), pp. 178–181.
13. Graham and Hays, *Managing the Public Organization,* p. 153.
14. *Ibid.,* pp. 153.
15. Kurt Lewin, Ronald Lippit, and Ralph White, "Patterns of Aggressive Behavior in Experimentally Created Social Climates," *Journal of Social Psychology,* May 10, 1939, pp. 271–301. Also reviewed in Cole Blease Graham, Jr., and Steven W. Hays, *Managing the Public Organization* (Washington, DC: Congressional Quarterly, Inc., 1986).
16. Alan V. Brunacini, *Fire Command* (Quincy, MA: National Fire Protection Association, 1985), p. 2.

17. Lussier, *Supervision,* p. 365.
18. Leslie W. Rue and Lloyd L. Byars, *Supervision Key Link to Productivity* (Homewood, IL: Irwin, 1990, ed.), p. 291.
19. Lussier, *Supervision,* p. 365 and Rue and Byars, *Supervision Key Link,* p. 291.
20. National Fire Academy, *Organizational Theory and Practice* (Emmitsburg, MD, 1991).
21. Rue and Byars, *Supervision Key Link,* p. 294.
22. Robert Blake and Anne A. McCanse, *Leadership Dilemmas—Grid Solutions* (Houston, TX: Gulf Publishing Co., 1991).
23. Cole Blease Graham, Jr., and Steven Hays, *Managing the Public Organization,* 2nd ed. (Washington, DC: Congressional Quarterly Press, 1993), pp. 183–185.
24. Lussier, *Supervision,* p. 367.
25. Rue and Byars, *Supervision Key Link,* p. 297.
26. Lussier, *Supervision,* p. 368.
27. National Fire Academy, *Instructor's Guide* (Emmitsburg, MD), pp. 8–17.
28. Ronny J. Coleman, *Residential Sprinkler Systems: Protecting Life and Property* (Quincy, MA: National Fire Protection Assn., 1991), p. 8.
29. Robert Tannebaum and Warren Schmidt, "How to Choose a Leadership Pattern," *Harvard Business Review,* May 1973.
30. Harry R. Carter and Erwin Rausch, *Management in the Fire Service,* 2nd ed. (Quincy, MA: National Fire Protection Assn., 1989), pp. 58–59.
31. Graham and Hays, *Managing the Public Organization,* p. 182.
32. Lussier, *Supervision,* p. 373.
33. For further discussion of Deming's work and influence, see the following: Lloyd Dobyns and Claire Crawford-Mason, *Quality or Else: The Revolution in World Business* (Boston: Houghton-Mifflin, 1991); W. Edwards Deming, *Out of Crisis* (Cambridge, MA: MIT Press, 1986); Kenneth Ebel, *Achieving Excellence in Business: A Practical Guide to the Total Quality Transformation Process* (New York: Dekker, 1991); and Rafael Aguayo, *Dr. Deming: The American Who Taught the Japanese about Quality* (New York: Simon & Schuster, 1990).
34. Graham and Hays, *Managing the Public Organization,* p. 188.
35. Op. cit.
36. Noel M. Tichy and Mary Anne Devanna, *The Transformational Leader* (New York: John Wiley & Sons, 1986), p. 4.
37. Lussier, *Supervision,* p. 374.
38. Thomas J. Peters and Robert H. Waterman, Jr., *In Search of Excellence: Lessons from America's Best-Run Companies* (New York: Harper and Row, 1987) and Thomas J. Peters, *Thriving on Chaos: Handbook for Management Revolution* (New York: Alfred A. Knopf, 1987).
39. William B. Johnston and Arnold E. Packer, et al., *Workforce 2000: Work and Workers for the 21st Century* (Indianapolis, IN: Hudson Institute; Washington, DC: U.S. Department of Labor, 1987).
40. Marilyn Loden and Judy B. Rosener, *Workforce America! Managing Employee Diversity as a Vital Resource* (Homewood, IL: Business One Irwin, 1991), pp. 182–194.

41. The behavioral style inventory presented here was developed and copyrighted by Dr. N. Ted Oravitz of Performance Development Associates of San Antonio, TX. Permission to use this inventory or other associated components must be secured from Dr. Oravitz.

REFERENCES

Aquayo, Rafael. 1990. *Dr. Demming: The American Who Taught the Japanese About Quality.* New York: Simon & Schuster.

Bennis, Warren, and Burt Nanus. 1985. *Leaders: The Strategies for Taking Charge.* New York: Harper & Row.

Berkley, George, John Rouse, and Ray Begovich. 1991. 5th ed., *The Craft of Public Administration.* Dubuque, IA: Wm. C. Brown.

Blake, Robert, and Anne A. McCanse. 1991. *Leadership Dilemmas—Grid Solutions.* Houston: Gulf Publishing Company.

Brunacini, Alan V. 1985. *Fire Command.* Quincy, MA: National Fire Protection Association.

Carter, Harry R., and Erwin Rausch. 1989. 2nd ed., *Management in the Fire Service.* Quincy, MA: National Fire Protection Association.

Coleman, Ronny J. 1991. *Residential Sprinkler Systems: Protecting Life and Property.* Quincy, MA: National Fire Protection Association.

Fox, Douglas M. 1979. *Managing the Public's Interest.* New York: Holt, Rinehart and Winston.

Graham, Cole Blease Jr., and Steven W. Hays. 1986. *Managing the Public Organization.* Washington, DC: Congressional Quarterly, Inc.

Graham, Cole Blease Jr., and Steven W. Hays. 1993. 2nd ed., *Managing the Public Organization.* Washington, DC: Congressional Quarterly Press.

Henry, Nicholas. 1991. 3rd ed., *Doing Public Administration Exercises in Public Administration.* Dubuque, IA: Wm. C. Brown.

Howell, Jane M., and Bruce J. Avolio. May 1992. Vol. VI, Number 2. "The Ethics of Charismatic Leadership: Submission or Liberation." *The Executive. Denver: Academy of Management.* pp. 43–54.

Johnston, William B., and Arnold E. Packer, et al. 1987. *Workforce 2000: Work and Workers for the 21st Century.* Indianapolis, IN: Hudson Institute; U.S. Department of Labor.

Loden, Marilyn, and Judy B. Rosener. 1991. *Workforce America! Managing Employee Diversity as a Vital Resource.* Homewood, IL: Business One Irwin.

Lussier, Robert N. 1989. *Supervision: A Skill Building Approach.* Homewood, IL: Irwin.

National Fire Academy. 1991. *Organizational Theory and Practice.* Emmitsburg, MD.

Oravitz, N. Ted. San Antonio, TX.

Ouchi, William. 1981. *Theory Z: How American Business Can Meet the Japanese Challenge.* Reading, MA: Addison-Wesley.

Peters, Thomas J., and Robert H. Waterman, Jr. 1982. *In Search of Excellence: Lessons from America's Best-Run Companies.* New York: Harper and Row.

Peters, Thomas J. 1987. *Thriving on Chaos: Handbook for a Management Revolution.* New York: Alfred A. Knopf.

Rue, Leslie W., and Lloyd L. Byars. 1990. 3rd ed., *Supervision Key Link to Productivity.* Homewood, IL: Irwin.

Thompson, Justin J. 1992. "Recruitment and Retention of Volunteer Fire Fighters." presented at the Annual Convention of the National Fire Protection Association. New Orleans, LA. May 18, 1992.

Tichy, Noel M., and Mary Anne Devanna. 1986. *The Transformational Leader.* New York: John Wiley & Sons.

Yukl, Gary A. 1981. *Leadership in Organizations.* Englewood Cliffs, NJ: Prentice-Hall, Inc.

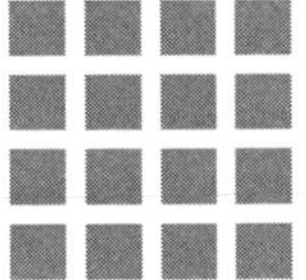

CHAPTER 3

Cultural Differences in the Community

Fire departments must not lose sight of the meaning behind the designation "fire service." A fire department does in fact provide services to its community. Like other service organizations, fire departments should consider members of the community as clients or customers to reinforce the concept of public service. Recognizing its public purpose, the fire service can more readily adapt the service to changes in the community.

When a business learns that the composition of its market or customer base has changed, it will modify both its products and its marketing plans to fit the needs of its customers. For example, the increase in the number of individuals over age 55 who travel has led many hotel chains to offer special "senior savings" packages to attract this market segment. The increase in the number of Hispanics in the U.S. has given rise to Spanish-language television and radio stations. This has been accompanied by advertisements in Spanish. The fire service must also begin to target its services toward the changing characteristics of its clientele. Indeed, some fire departments have already begun to respond.

What kinds of changes will be necessary? There are a number of national trends that the fire service should recognize. These provide a basis for demographic generalizations and can identify specific minority populations that will benefit from targeted fire prevention programs. In addition, local fire departments must evaluate changes taking place in their communities in order to be prepared to serve their clients appropriately. National trends provide generalities; local assessments offer specific information. This chapter will look at the national trends, discuss how to identify local changes, and then offer suggestions on how to adapt to the changes taking place.

NATIONAL TRENDS

The demographic composition of the U.S. is being reshaped by two major facts: immigration and birth rate. The U.S. was built on immigration and has maintained a reputation of being a refuge and place of opportunity. The countries of origin of immigrants have changed over the years and are increasingly diverse. According to 1990 population figures, the U.S. continues to take in more immigrants than any other nation.[1]

The rate of immigration continues to grow. More immigrants, 8 to 10 million, are projected to enter this country in the 1990s than in the 1980s, when 7 million arrived. [See Figures 3-1(a) and 3-1(b).] These

FIGURE 3-1 **(a)** Number of foreign-born residents in the U.S., current and projected. **(b)** Approximately one-fifth of the nation's total population by the year 2010 will be immigrants or children of immigrants.

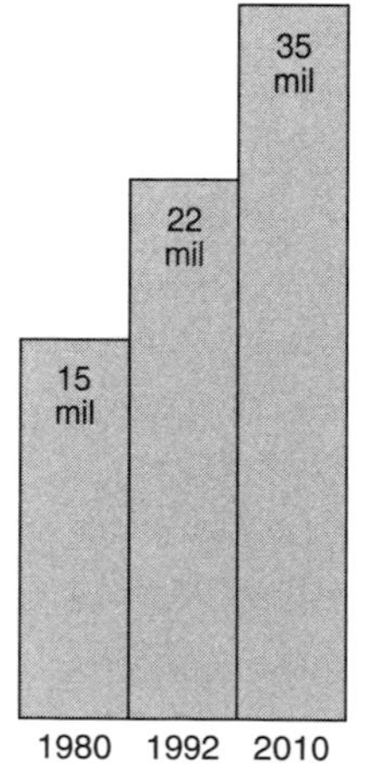

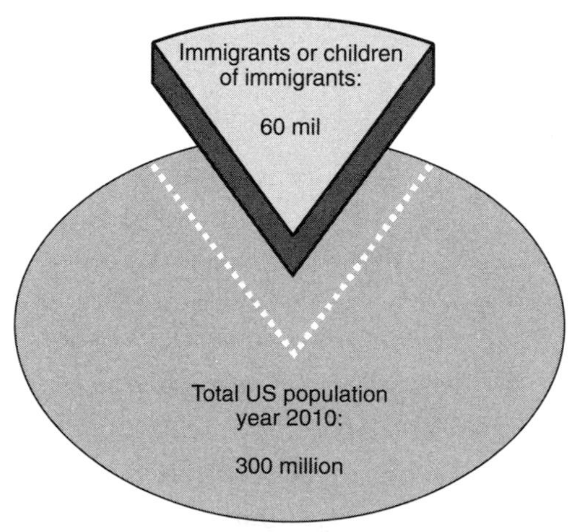

new residents will come primarily from Mexico, the Philippines, Vietnam, Korea, India, China, the Dominican Republic, Jamaica, El Salvador, Iran, Laos, Taiwan, eastern Europe, Ireland, Great Britain, and Columbia. Initially these immigrants head predominantly for California, New York, Texas, Florida, Illinois, and New Jersey. However, the immigrants and their children continue to move to all areas of the country in search of economic opportunity.

Many of today's new entrepreneurs are immigrants. For example, more than 10 percent of the Koreans who came to the U.S. in the 1980s now own and operate small businesses.[2] When making a company inspection or a detailed fire and building safety code inspection in small businesses, fire officers must deal with different cultures and find new ways to explain safety codes and regulations.

In addition, more of the workers in major businesses are immigrants, and the numbers are continuing to increase. These individuals do not have the same educational background in fire safety and awareness that results from fire prevention efforts in U.S. schools. The concern for safety in the workplace is not as great in many cultures as it is in the U.S. Thus, immigrant workers may not be aware of many of the safety standards that must be met or of the environmental dangers that are present on the job site. As a result, fire codes may be violated by business people or workers who aren't aware of fire safety or other requirements when making decisions.

New immigrant populations often retain and use their native language while assimilating into the culture of the U.S. During this transition period, serving communities with non-English-speaking populations and differing cultural norms can challenge the fire service. The most obvious challenge is dealing with the language differences. How does a fire department respond to a 911 call when the caller cannot be understood? Do members of the community know to call 911 for help?

Outreach and education programs must be refocused to be effective. Although cities have been the historic centers of multiculturalism, diverse populations are found in suburban and rural America as well. Increasingly, our entire society is becoming multicultural.[3] Fire departments need to know their communities, understand their cultural makeup, and be aware of changes taking place.

KNOWING YOUR COMMUNITY

Fire officers need to stay current with changes in the population of their service areas. It is generally known that an aging population will place an increased demand on EMS services. However, shifts in ethnic and cultural composition of an area can also place different service demands on the fire department. In order to meet these challenges, fire officers must stay ahead on the learning curve, planning to meet the challenges of the changing population.

The fire chief or designated officer should meet regularly with the planning director or a member of the municipal planning agency. This department generally has the most recent census data as well as demographic data from other study sources, and can provide a fairly accurate picture of population changes taking place in the community. In smaller areas, it may be necessary to go to county or state agencies for such information. Obviously, the smaller the population served, the easier it is for the fire department to be aware of changes in the community makeup. This makes it easier for the department to observe change, but in no way lessens the importance of *using* the information to best serve the community.

Another source of information on changes in the community is the state department of social services or local social service agencies. These departments frequently have direct contact with new immigrants, minority ethnic groups, and senior citizens. Working with these departments and agencies can increase the fire department's awareness of the community as well as enhance the understanding of fire department operations and concerns among other members of the local government. (See Chapter 7, Interacting with Other Public Administrators, for further discussion of this.)

Leaders of local religious organizations can also assist the fire department in identifying the presence of new minorities and understanding their culture. Local religious groups often sponsor programs that work with new immigrants to help them find housing and employment. In addition, the presence of new places of worship ministering to a larger or growing group of immigrants is a clear indication that the ethnic composition of the community is changing.

FIGURE 3-2 Public education programs provide an opportunity for the fire department to interact with members of the community.

Fire departments involved in public education programs have an additional source of primary information concerning changes in the composition of the local population. A plurality of the school-age population in California is Hispanic. Fire officers conducting fire education programs in the public schools may notice changes in the demographic composition of the children attending the programs, some of whom may not use English as their first language. Likewise, the fire inspection bureau should notice changes in ownership of small businesses.

Public school administrators can usually assist the fire department in identifying how many students from different cultures are attending public schools. The school systems maintain records of which students speak a language other than English as their first language. This information, especially when tracked over time, can identify trends as well as the current status of changes in a community's ethnic composition. Noticing these changes is not enough, however. The trends must be formally noted and addressed. The department must adapt its programs to new populations in order to remain effective.

Meeting the Needs of the "New" Community

Fire department personnel must become aware of the cultural attitudes and practices of its clientele. Language difference is the most obvious immediate challenge facing the fire service. In larger cities, such as Dallas, Texas, more than thirty-two separate languages have been identified as the primary language spoken in the home. While children generally learn English in the schools, the parents—and especially grandparents—may never learn. Language barriers make communicating during an emergency very difficult. It is not necessary for the fire department to have thirty-two translators on staff; however, it is critical that fire officers know where to get help with translation if it is necessary. And this assistance must be available quickly. This is where cooperation with other local departments and agencies can prove beneficial.

Fire prevention literature often finds its way into the homes of immigrants by way of the children who attend a program at school. Obviously, if the material is presented in both the native language and English,

FIGURE 3-3 EMS calls are handled efficiently when the dispatcher understands the caller's needs. Multilingual dispatchers are often essential in communities with large minority populations.

FIGURE 3-4 Fire safety literature can be introduced to immigrant families through school-based fire education programs.

the impact in the home will be far greater. In multilingual communities it is a good idea to add a "take the news home" assignment to the fire prevention program. In this way parents become informed about fire safety and local emergency procedures.

Children often learn at a young age to serve as interpreters for non-English-speaking members of their families. They can assist their families in times of emergency—if they have been trained in advance. Although it is a challenge, it is important for fire departments to expend the effort to reach out to these new members of the community and help them prepare for an emergency.

Along with language barriers, cultural misunderstandings can be barriers to effective communication about safety. Fire departments serving communities with large, diverse populations need to learn about the specific culture of each group. For example, in some Asian cultures cooking over small charcoal stoves is the norm. Families may cook over a hibachi grill indoors. The danger of fire is obvious. However, to simply tell these families that they cannot do this because it is against the law is not an effective approach. It is likely that such instructions will be ignored because they contradict a cultural norm. The fire department may alienate a segment of the community by being insensitive and unaware of an effective means of communicating safer alternatives.

Thus, in the previous example, a fire officer, accompanied by an interpreter, might approach the families with a general presentation on fire safety and then focus on the particular danger of using grills indoors. The fire officer might then ask for the families' help in avoiding such dangers and thank them for their cooperation. The dangers are identified and the families maintain their honor.

In order to develop effective fire prevention education programs, the fire service must learn about and understand the cultures of different groups. In this manner, they may be able to identify fire hazards common to daily living. Further, the fire service must understand how to effectively communicate fire safety practices to various segments of the population. This requires sensitivity and awareness on the part of the fire department. In order to teach the community what the fire service is, how it operates, and how to increase fire safety, the fire service must first learn about those whom it serves. This is called cultural awareness.

CULTURAL AWARENESS

Cultural awareness is a growing area of knowledge that is receiving considerable attention and increased study. Organizations in both the public and private sectors are recognizing the importance of being aware of the culture of both their clientele and members. **Culture** is defined as a system of customs and traditions, sometimes coupled with beliefs, that influences or drives daily activities and methods of interaction.

Culture provides an orientation to life. It is a way of looking at the world as well as the compilation of various activities that are customary to a certain group of people. Increasing awareness of multiple cultures encourages greater sensitivity and more perceptive observations of people from different cultures. Greater understanding of situations involving diverse cultural groups allows for more accurate and effective response. Cultural awareness can improve both client and employee relations, as it creates an understanding of the reasons for different behaviors.

In order to effectively communicate fire safety and to interact positively with culturally diverse clients, fire service personnel must consider the underlying values of a culture. These existing values influence thought and action and must be understood and appreciated. All of us have a value system that influences our actions and perceptions. It is important to be aware of our individual biases, which are based on our own value systems. (See Chapter 12, Ethics.) Failing to recognize that value systems based in other cultures are simply different from one's own will hamper an individual's ability to deal with and influence members of that culture. In a service where changing behavior and attitude are important, where we are trying to instill life safety awareness and practices, cultural biases can sabotage our efforts.

A number of skills are common to various cultural awareness training programs. The following skills are important for both managing different cultures and transferring knowledge to a different culture.

Demonstrate respect. When working with people of different cultures, it is important to demonstrate respect for that culture. This needs to be done both verbally and nonverbally. Thus, it is necessary to be familiar with cultural taboos and insults that may differ from one's own. In some cultures asking about a person's ancestors, which may arise when compiling a medical history, is considered impolite and a violation of family privacy. In this situation, questions must be asked differently.

Be nonjudgmental. Avoid evaluating or being moralistic about different ways of life. Explain that certain activities are safe or unsafe rather than "right" or "wrong." Deal with the facts and results of activities and do not put yourself in a position of superiority.

Show empathy. Try to understand a situation from the other's point of view. Try to see things as members of the minority community see them. This will help you develop a better perspective of why their attitudes and actions are as they are. Once this is clear, you can then begin to develop ways of changing unsafe practices within the boundaries of their culture.

Be flexible. Accept the fact that in order to achieve the same goal, you may have to follow different procedures in different communities. For example, while the presence of a red fire truck and offerings of free literature may attract members of one cultural community, it may alienate members of another. Some communities will participate in activities on Saturday, others will not. In fire prevention education activities we have discovered that different ways of communicating messages work better with different audiences; the same is true for different cultures.

Tolerate ambiguity. Do not become upset with differences. Realize that some frustration is inevitable, but that it should not lead to force or assumption of superiority. Fire officers and fire fighters must continue to deal positively with changing circumstances and different people.

These skills are helpful for managers in general, and are crucial when dealing with multiple cultures. Applying these ground rules for dealing with different segments of the community will enhance the effectiveness of the delivery of fire department services. In addition, it will increase community support of the fire department as residents come to recognize the fire department's consideration of their cultural values.

There are two fundamental principles to remember when facing cultural differences and trying to achieve cultural awareness:

1. "Different" does not mean "good" or "bad" or "better" or "worse." Different means not the same, distinctive, unique, diverse.
2. Learning about the culture of members of the community enables one to develop as an individual and to better serve the community as a professional.

Cultural Awareness in Fire Inspections

Tradition is the root of many fire service activities. Some activities take place just because "that's the way it's always been done." Given this, members of the fire service should have a greater understanding of those individuals from cultures who respond to questions concerning unsafe practices by saying, "But that's the way we have always done it—we have always cooked our meals in the living room next to the table, and charcoal makes the food taste better." Just as members of the fire service expect tolerance and understanding of its operating tradition from society, fire fighters and fire officers must be tolerant and understanding of people faced with challenges to their traditions.

One of the first ways to address changes in tradition is to gain an understanding of why those traditions exist–what started the tradition? For example, a fire department in the Midwest washed the wheel wells of each apparatus every day at 4:30 p.m. It was a department tradition. The wells were washed whether the apparatus had left the station or not. However, the constant washing was wearing off the undercoating and leading to faster rusting of the wells. In trying to change this practice, officers met with resistance to changing routine and "tradition." One officer took the initiative to find out how this practice started. She discovered that when the department was first organized and used horses to draw the apparatus to the scene, the horse stalls were cleaned out and washed down at 4:30 p.m. each day. The horses then received their second feeding of the day. When motorized vehicles replaced horses, the routine was adapted, but the schedule remained the same. Once this was understood, the members of the department realized that the need for the traditional activity had changed over time. They were then more willing to alter the station routine.

In the same way, if fire departments wish to change the cultural patterns and traditional activities of people from various backgrounds, they must gain an understanding of *why* these traditions exist. Before simply citing business owners for code violations and *telling* them to do certain things, fire officers should find out why they follow the current practices, then explain why the code requires certain arrangements and explain the safety factors involved. In doing this, the officer should try to address the underlying reasons the owner/manager has for the current setup. Providing a frame of reference can achieve more voluntary compliance, which is more likely to endure.

Although this approach initially requires more time and initiative from the fire inspector, the results are more positive and enduring. Citations and compliance under threat often result in grudging compliance with the law–and no more. The intent of the law is lost, and respect for the law and its enforcers–the fire department–is shattered. In these instances, repeat violations often occur. The amount of effort expended by the fire department and the level of frustration experienced by both the members of the department and the community over time more than exceeds that expended with a culturally sensitive approach.

Where to Find Cultural Awareness Training

Cultural awareness and sensitivity training can be offered by national training groups or consultants or by locally based sources. There are a number of very good national cultural awareness and cultural sensitivity programs. However, the *Chronicle of Higher Education* reported that the trend of promoting ethnic and racial diversity has produced a rash of workshops and presentations that are hastily designed and of questionable value.[4] Other presenters have caused more division instead of understanding. Always check references before using an outside consultant.

The best sources for learning about different cultures are those located in your own community. Local presenters can reference local changes and concerns. Examples can be drawn from familiar areas, and fire fighters and officers can relate directly to the information being shared. Many national trainers do not customize their presentation to address the characteristics of a particular community. In addition, by using local sources, the fire department can use the training experience to build bridges of understanding, cooperation, and support within the community. There are a number of local sources that may be able to assist the fire department.

Colleges and universities in the area often have programs or faculty who work with the community and are able to present a program on cultural differences. Some universities have programs geared toward multicultural awareness and understanding. Others have certificate or degree programs in minority cultural areas such as Afro-American Studies, Hispanic Studies, or Asian Studies. Even when the institution does not have a program focused directly on the issue of cultural differences, there are generally faculty in departments of sociology, urban studies, political science, anthropology, or even history who can assist.

Ethnic community groups often have cultural awareness programs that they share with members of the community. Even if they do not have a formal program, ethnic community groups can be extremely helpful because they are typically organized to preserve their ethnic heritage. This means they are familiar with the customs and traditions of their culture and can explain their origin. This is precisely the type of information the fire department needs in order to better relate to the community. In addition, because these groups are based in the community, they can serve as a conduit to take fire safety information from the department back to the community.

Social service agencies that work with immigrant groups have an understanding of cultural differences as well. An administrator in the agency can share information on successful and unsuccessful outreach programs the agency has tried. The fire department can learn from these experiences. It may also be possible to have someone from the agency give a presentation to department members or identify someone in the community who could be of assistance.

Religious diversity is part of cultural diversity. **Places of worship** in minority neighborhoods are a good source of information. They can provide information about the people and their customs and provide a means for public service personnel to meet members of the community. Places of worship are also gathering places, and their leaders are often dedicated to helping the community in any way possible. As such, they are generally receptive to assisting fire departments and other public agencies in establishing working relationships and fostering understanding between the community and the public agencies. It is important to be familiar with the various roles religious leaders play in other cultures to establish these relationships.

These are only some of the sources for information on immigrant communities. **Public schools** can be helpful as can local **community centers.** Look within your community for assistance first. Then, if you decide you need professional training, you will have a better understanding of the needs of the department and will be better prepared to screen and select the most appropriate training program offered.

SUMMARY

Cultural awareness and sensitivity are necessary in order to serve the community. Members of the fire department must understand the nature of the community they serve. They must become aware of cultural differences and the reasons for varying actions and attitudes.

The most important element of cultural awareness is the acknowledgment that no culture is inherently superior. In order to integrate a culturally diverse organization and society, individual and cultural characteristics must be respected.

The fire service is recognizing the importance of the word "service" in its name. In order to serve the community, the fire service has to adjust to the changing needs of the population. While immigrants may learn English and adapt their traditions over time, they will not adapt overnight. The fire service has to learn how to work with diverse populations if it is going to be successful in protecting the lives and property of all residents.

CASE STUDIES

Case #1

The Combination Fire Department responds to a residential structure fire in a two-family dwelling unit located within 1/4 mile of an unstaffed fire station.

Two lives are lost, including those of a mother and her 2-year-old son. Three injuries are incurred, involving the grandmother of the deceased boy from smoke inhalation, one other child, and one fire fighter when a wall collapses.

Notification of the fire is delayed because neither of the dwelling units have smoke detectors installed as required by the local fire code. Once the fire is discovered, the occupants are not sure (or did not understand) what action to take. This combination of events causes an 11-minute delay in notifying the fire department.

Upon arrival, the first engine company notes that the fire is confined to the dwelling unit on the east side and that flames are shooting from a large front window; heavy smoke is discharging from several windows on the second floor. Although the officer in charge makes an effort to speak with several people at the scene concerning other occupants who might still be in the structure or be trapped, the attempt is futile since no one is able to speak or understand English, and they cannot communicate with the officer. The officer has no idea what language the people are speaking, and other onlookers are also unable to communicate with the apparent residents.

Entering through the rear of the home, two fire fighters attempt to search for possible occupants. One fire fighter locates a small child and tries to remove her when a wall collapses, injuring them both. The second fire fighter is able to assist them out of the structure. After that incident, no more rescue attempts are made. The deaths are not discovered until after the fire is extinguished. Each of the units involved in the fire was occupied by a Cambodian refugee family.

During the critical incident debriefing, one of the fire fighters who responded with the third alarm companies relates that his church had just sponsored five Cambodian families. These families had arrived a month earlier, and most of the family members did not speak English. In fact, he was assisting with some special English classes held at the church on Saturday afternoon. The church had arranged for housing and had found jobs for one or two members of each family.

1. How could this incident have been prevented? Inspections? Information? Training? (Of whom? How?)
2. How could the fire department have been better prepared to handle the incident?
3. What action(s) should the fire department take now?

Case #2

Last year a major food processing company purchased large areas of farmland to grow tomatoes for food processing. During harvest season, Spanish-speaking migrant workers come into the area to pick tomatoes. A few weeks after harvesting begins, the Volunteer Fire

Department begins to experience an increase in EMS calls and grass fires.

Response on EMS calls has been problematic because the injured party cannot always speak English. Although a few members of the department studied Spanish in high school, none of the EMTs speak Spanish well enough to communicate with the patient. In a couple of serious incidents, the squad members decide to transport a fellow migrant worker to serve as translator.

In one case, late notification and misunderstanding of the symptoms almost resulted in tragedy. In other situations, patients requiring only minor first aid that could be administered on site have been needlessly transported. The frustration felt by fire department members is adversely affecting morale. Some members are hesitant to respond, fearing that the call may be complicated by language difficulties. This is placing an unfair burden on other members, who are rapidly approaching burnout.

The grass fires appear to have two major causes: campfires and cigarettes. Although there are some cooking sites and alcohol stoves available in the migrant work camp, some of the workers use campfires for heating water to wash clothes, clean, etc. In addition, sitting around a campfire in the evening is a popular activity because the electricity sometimes goes out and lamps are not always available. Many of the migrant workers smoke, and due to the outdoor nature of their work and living environment, cigarettes are sometimes discarded by simply "pitching" the butt.

If the growing and processing operation is a success, the company plans to move a satellite processing plant to the community and hire more locals in permanent positions. This would help offset the layoffs anticipated as a result of the downsizing of other manufacturing plants in the area. The company has been holding talks with the mayor and other community leaders, including school principals, in an effort to provide needed services to the migrant camp without placing too great a burden on the community. At the same time, the company managers have made it clear that they recognize the benefit the operation will have to the town and they are not willing to pay unreasonably for additional services. The company managers are more interested in a partnership approach, where each contributes to the solution of a mutual problem.

1. What are the major problems facing the fire department? (Morale, resources, burnout, communication, etc.?)
2. How could the fire department prevent some of the grass fires? What would it take to do so? Who else would have to be involved in the effort?
3. How could the EMS situation be improved?
4. How should the fire department approach the solution to these difficulties?
5. What should the fire chief do? Who should he talk to? In what order? Why?
6. Who should be responsible for the increase in local services associated with the workers hired by the food processing company?
7. What are some alternative solutions?

Case #3

For the past 7 years, Captain Mikee, the Combination Fire Department's chief fire safety inspector, has worked closely with both the state fire marshal and the safety officials at the state university to maintain an adequate level of safety at the campus, located in the department's response district. Since the chemistry department is the largest academic unit on campus, special emphasis has been placed on the safe handling and storage of hazardous chemicals and substances. All of the materials have been reported to the Local Emergency Planning Committee under SARA Title III. The regional hazardous materials response team developed an emergency response plan and has trained to the plan by conducting two simulated spill emergencies within the past 18 months.

All buildings on campus are inspected semi-annually by Mikee, and any fire safety deficiencies are reported to the university safety officer at once. Because the use and storage of hazardous chemicals poses a real danger to students and professors who use the chemistry department facilities, Captain Mikee has recommended that the university designate a faculty person as chemistry safety officer for just that department. In addition, he suggested that all personnel receive 4 hours of hazardous materials awareness training at the beginning of each academic year.

For the past 4 years, Dr. King has functioned as the chemistry safety officer and worked closely with the fire department during the preplanning and training development phases. At the end of last semester, Dr. King retired and was replaced by Dr. Pradhaumme, a new faculty person who was born and educated in a Middle Eastern country. Captain Mikee decides to visit the department, introduce himself, and inform Dr. Pradhaumme of the cooperative arrangement the fire department maintains with the university's safety department. Dr. Pradhaumme is accommodating and assures Mikee that safety will continue as the top priority in the chemistry department.

The following month, Mikee schedules a follow-up inspection of the chemistry department to observe how the new chemical safety officer is handling the job. Upon entering the building, he is appalled to discover a large number of compressed gas cylinders standing freely along the walls of the main corridor and two graduate students moving a flammable compressed gas cylinder from a cart to the wall area. Mikee immediately orders the action stopped and proceeds to Dr. Pradhaumme's office. He explains that the main corridor is considered a means of egress by NFPA *101*®, *Life Safety Code*®, and that storage of compressed gas cylinders in the corridor violates the code and creates a dangerous condition. Mikee also reminds the professor that NFPA 58, *Standard for the Storage and Handling of Liquefied Petroleum Gases,* prohibits the storage of cylinders in unapproved spaces.

When confronted with these issues, Dr. Pradhaumme appears puzzled as he explains that he is unaware of such fire and safety regulations. He states that in his own country and at his former university, it was common practice to store compressed gas cylinders near laboratories in which they were to be used, instead of at one central location that may be inconvenient for some researchers. That way, replacement of a compressed gas cylinder could be accomplished with little time lost. Besides, the professors and students in the chemistry department are acutely aware of the dangers of compressed gas and would not under any circumstances disturb any of the cylinders in the hall.

1. Why was Dr. Pradhaumme unaware of safety regulations?
2. What should Captain Mikee do now?
3. Who else should become involved in the situation?
4. Is Dr. Pradhaumme's faith in the students well founded? Why or why not?
5. How might the fire department ensure that this not happen again?

Case #4

The Combination Fire Department is serving a growing number of Spanish-speaking people in the community. A number of Mexican-Americans has applied to join the department, but there are no current openings. They are seeking full-time positions with the department, but the current hiring procedure is to hire all full-time personnel from the volunteer ranks. At this time, the department is not even accepting volunteer applications. However, because of the language barrier, Chief Marks wants to make an exception and hire a Mexican-American staff person in the fire prevention division.

The volunteers are upset about this proposal because they feel they have the right to full-time positions first. This is causing hard feelings between the fire department, especially volunteer members, and the Spanish-speaking community.

1. What should Chief Marks do?
2. Was he right to consider hiring a Mexican-American full-time?
   - (a) What are the positive and negative aspects of each alternative?
   - (b) Should he have let that possibility become public? Why or why not?
3. How might Chief Marks prevent such situations in the future?

ENDNOTES

1. U.S. Department of the Census, *Analysis of Population Trends, 1990* (Washington, DC: Government Printing Office, 1992).
2. "Kiplinger Newsletter," July 1992, and U.S. Census of Population and U.S. Census of Business and Industry (Washington, DC: Government Printing Office, 1992).
3. U.S. Department of the Census, *U.S. Census of Population, 1990* (Washington, DC: Government Printing Office, 1992).
4. "So-Called Experts on Cultural Diversity Need to Be Subjected to Careful Scrutiny," in *Chronicle of Higher Education,* Vol. 38 Iss. 12, Nov. 13 1991, pp. B2–B3.

REFERENCES

Allen, James P. "Where Diversity Reigns." *American Demographics.* Vol. 12, Iss. 8, August. pp. 34–38.

Armstrong, Liz Schevtchuk. 1991. "Census Confirms Remarkable Shifts in Ethnic Makeup." *Education Week.* Vol. 10, Issue 26, March 20. pp. 1, 16.

Buenker, Hohn D., and Lorman A. Ratner, eds. 1992. *Multiculturalism in the United States: A Comparative Guide to Acculturation and Ethnicity.* Westport, CT: Greenwood Press.

Denton, Nancy A. 1991. "Patterns of Neighborhood Transition in a Multiethnic World." *Demography.* Vol. 28, Iss. 1, February. pp. 41–63.

Gardenswartz, Lee, and Anita Rowe. 1989. "The Multi-Cultural Workforce." *Working World.* May 16, p. 16; June 26, p. 34; August 14, p. 20.

Johnston, William B., and Arnold E. Packer. 1987. *Workforce 2000: Work and Workers for the Twenty-first Century.* Indianapolis, IN: Hudson Institute, Produced for the U.S. Department of Labor.

Kern, Marilyn. 1989. "Minorities 2000." *Public Relations Journal.* Vol. 45, Iss. 8, August. pp. 14–18, 21–22.

Kitano, Harry L., and Roger Daniels. 1988. *Asian Americans: Emerging Minorities.* Englewood Cliffs, NJ: Prentice Hall.

Lambert, Wallace E., and Donald M. Taylor. 1990. *Coping with Cultural and Racial Diversity in Urban America.* New York: Praeger. 292 pp.

Laporte, Suzanne B. 1991. "Cultural Diversity: The Sting of the Subtle Snub." *Working Woman.* Vol. 16, Iss. 1, January. pp. 53–55.

Monroe, Linda R. 1989. "Culture Shock Hits Health Care." *Los Angeles Times.* August 19, Part I, p. 1.

Montville, Joseph V., ed. 1990. *Conflict and Peacemaking in Multi-Ethnic Societies.* New York: Lexington Books.

Morrison, Ann M., and Mary Ann Von Glinow. 1990. "Women and Minorities in Management." *American Psychologist.* February. pp. 200–208.

Njeri, Itabari. 1989. "Intercultural Etiquette: When Different Groups Converge, the

Ignorant and the Curious Can Be Unexpectedly Rude." *Los Angeles Times.* April 2, Part IV, p. 1.

Van Horne, Watson A. 1988. *Ethnicity and Health.* Institute on Race and Ethnicity.

Wilkerson, Isabel. 1989. "Study Finds Segregation in Cities Worse than Scientists Imagined." *New York Times.* August 5. p. 6.

CHAPTER 4

Human Resources Management

The ability to motivate, lead, and relate to different individuals in the fire service is a vital aspect of human resources management. From this perspective, human resources management is a recurring topic throughout this book. This chapter focuses on the legal requirements involved in administering and managing fire service operations and employment practices. It also addresses the challenge of meeting federally mandated employee protection and equal opportunity requirements.

The fire service poses some unique human resources challenges and shares a number of concerns and challenges with other areas of public administration. This chapter points out both the similarities and differences, and concentrates most attention on those aspects of immediate concern and relevance to the fire service.

The Fair Labor Standards Act regulates hours, wages, and other working conditions. The development of the various requirements and the special regulations covering operations of fire departments are presented to provide an overview of legal requirements. When fire officers have specific legal questions regarding fair labor practices, it is always advisable to seek legal counsel. This discussion provides an overview and a sense of the intent of the legislation. It does not present detailed interpretation of the law.

The Occupational Safety and Health Administration (OSHA) is charged with protecting the health and safety of employees in the workplace. The agency sets standards for employers to follow concerning training, information, and protective actions against occupational hazards to employees. Those OSHA regulations of most relevance to the fire service will be discussed.

Workers' compensation is a special area of fair labor practice that has tremendous impact on the operating costs of fire departments. We will consider the requirements, options, and impact on the total operating budget of the fire department. Increased operating costs in the fire department impact the funds available for other local government operations, except where fire districts are the taxing entities.

The fire service mirrors the society it serves. As society becomes more multicultural and as women continue to enter and stay in the workforce, more members of minority cultures will join the fire service. It is important for fire officers to be aware of this change and to deal with cultural differences in the organization. Valuing diversity in the fire department will enhance the overall activities and contributions of the department. This is not easy, however. It demands awareness and sensitivity.

Among the many laws governing equal compensation and equal employment are the Equal Pay Act, the Civil Rights Act, and the Age Discrimination in Employment Act, which we will explore within the context of the fire service.

Equal employment opportunity and affirmative action are increasingly important for the fire service as the demographic composition of the nation and the fire service changes. The fire service must consider these requirements along with the problems associated with sexual harassment in its daily operations.

The discussion of legal mandates is followed by an overview of labor agreements, grievance procedures, retirement programs, substance abuse, and skills development programs. These aspects of human resources management influence morale and work performance, and they require some specialized knowledge. In addition, the budgetary impact of a policy must be considered when determining whether or not to implement a program and what type of program to consider.

LEGAL REQUIREMENTS

Fire officers must be familiar with a number of federal laws governing the employment and treatment of workers. Most of these came about as a result of unsafe practices. An example of this is permitting fire fighters who are not protected by self-contained breathing apparatus (SCBA) to enter into an emergency fire or hazardous materials situation where a hazardous substance may have been released. An example in the private sector is allowing rubber workers to use benzene, a suspected carcinogen, to remove residue from their hands. OSHA now requires that these workers be informed of the toxicity of the chemical, trained in its safe use, and provided with protective equipment through the OSHA Hazard Communication Program. Various OSHA regulations have withstood several court tests and will probably provide precedents to withstand future challenges.

Fair Labor Standards Act

The initial version of the Fair Labor Standards Act (FLSA) was passed in 1938 and is widely known as the "Wage and Hour Law." It is important to remember that once legislation has been passed by Congress, it stays in effect unless it is rescinded or amended. The original FLSA has been amended a number of times. Following is a list of the legislative amendments to the FLSA along with the relevant citations of public law. The FLSA legislation sets (1) the minimum wage rate, (2) the overtime pay standards, (3) the standard for straight time versus overtime, (4) maximum work hours, and (5) the child labor restrictions.

An overview of the chronological development of FLSA legislation and relevant judicial findings of greatest relevance to the fire service is presented below. Initially the FLSA did not apply to public employees. However, a series of court cases and legislative amendments has changed this. In 1966 amendments extended the coverage of FLSA to certain state and local government employees. The constitutionality of this extension was challenged and upheld in the 1968 Supreme Court Case *Maryland v. Wirtz.* Once the constitutionality of the application to public employees was established, Congress then expanded the FLSA in 1974 to cover all state and local government employees, except for certain identified employees—among them fire fighters. This exclusion was challenged in the case *Garcia v. San Antonio.* The Supreme Court ruled that the FLSA was fully applicable to all state and local government employees.

Legislative and Judicial History of the Fair Labor Standards Act Relevant to the Fire Service

1938 Congress enacted the Fair Labor Standards Act (FLSA), which established nationwide minimum wage and maximum hour standards for the first time. The 40-hour workweek and time-and a-half pay for all overtime hours applied to the private sector only. (52 Stat. 1060)

1966 Congress extended the FLSA to cover certain school, hospital, nursing home, and transit employees of state and local government. (80 Stat. 880)

1968 *Maryland v. Wirtz,* 392 U.S. 183, challenged the 1966 amendments on the basis of violation of the tenth amendment to the Constitution. The amendments were upheld by the Supreme Court, which concluded that the federal regulations were constitutional under the Commerce Clause and did not unduly interfere with the performance of governmental functions entrusted to the states.

1974 Congress expanded the FLSA to cover all state and local government employees except for a small number who were specifically exempted. These amendments included limited overtime exemptions for police officers, fire fighters, and related employees. [29 USC 207(k)] The work period could be identified as 7 to 28 days for the purpose of calculating compensation. Fire fighters could work a 53-hour workweek before overtime costs began to accrue. (88 Stat. 55)

1976 *National League of Cities v. Usery,* 426 U.S. 833. The Supreme Court overruled Wirtz and held that both the 1966 and 1974 amendments were unconstitutional to the extent that they interfered with the integral or traditional governmental functions of states and their political subdivisions. Schools and hospitals, fire protection, police protection, sanitation, public health, and parks and recreation were held to be traditional functions of state and local government.

1979 The Department of Labor issued the final regulations defining traditional and nontraditional functions of state and local government, adding libraries and museums and defining local mass transit systems and seven other functions as nontraditional. (93 Stat. 468)

1985 *Garcia v. San Antonio Metropolitan Transit Authority.* On February 19, the Supreme Court overruled *National League of Cities* and left the FLSA fully applicable to state and local governments.

1985 On November 6, Congress passed legislation defining the specific ramifications of *Garcia* for the fire service, as well as for other local government safety and transportation services. (99 Stat. 787)

The legislation passed in response to the *Garcia* ruling has special provisions for fire fighter compensation because of the unique nature of public safety positions (fire protection and law enforcement). (99 Stat. 787) The highlights of this legislation affecting the fire service follow. These points are important when monitoring schedules to achieve maximum performance and minimize overtime costs.

Summary of 1974 Amendments to FLSA Affecting Fire Service

- Time and a half in cash or compensation time for all overtime hours.
- Compensation time may be accrued and banked up to a maximum of 480 hours, pursuant to a collective bargaining agreement or, in the absence of a collective bargaining agreement, an understanding with the employee representative or individual employee. Cash must be paid after 480 hours of comp time has been accrued. All comp time must be awarded at the premium rate of time and a half. Comp time is a legal liability of the employer and, upon termination of employment, unused compensatory time shall be paid at a rate of compensation not less than:
  (a) the average regular rate received by such employee during the last 3 years of the employee's employment or
  (b) the final regular rate received by such employee, whichever is higher.

- Fire fighters may work special detail assignments without necessitating overtime pay, as long as the work is strictly voluntary on the part of the employee and the work is clearly being done for a separate and independent employer.
- Fire fighters may not perform volunteer work for the same agency they are employed by, or respond to volunteer calls in that same jurisdiction. Volunteers will not be deemed employees if they receive "nominal" payments for their services.

Highlights of the Current FLSA of Specific Concern to the Fire Service

Compensation for overtime can take the form of compensation time (comp time), which the fire fighter can take as time off, or as overtime pay at the rate of one and one-half times the standard hourly rate. Fire fighters may accrue up to 480 hours of compensatory time before they are required to take the time off or to receive monetary compensation. In addition, the employer must pay for any unused comp time upon termination of the employee, regardless of the cause of termination. Monitoring the amount of comp time being aggregated by department personnel is an important facet of managing department operating costs. Unmanaged accrual of comp time can be very costly for the department.

In the fire service, a standard workweek is defined as 53 hours. In addition, the amendments of 1974 also allow city government to set the standard work/pay period at 28 days to mitigate the impact of the legislation on fire and police departments. This arrangement offsets to some degree the controversy over paying fire fighters for sleeping. While this type of scheduling is done for insurance or safety reasons, it often remains a sore spot with public administrators during times of employee cutbacks. The *Public Administration Review* article titled "Pay for Time Not Worked" examined the average number of hours per week employees in three municipal departments were paid even though they were not working.[1] These periods of "not working" included coffee breaks, time at the water cooler, time spent eating when it was not done "off the clock," and, of course, sleeping. In the water department, 16 percent of the time paid was not worked; in the police department the amount of time climbed to 23 percent. The fire department numbers were staggering: 53 percent of the time a fire fighter is on duty, he or she is being paid for not working!

Studies like this, which reach a vast number of public administrators and public officials, prompt questions about the traditional 24/48 staffing pattern of the fire service. Under the 24/48 pattern an individual is assigned a 24-hour work shift, followed by 48 hours off duty, then another 24-hour work shift, etc. However, when a regular workweek is defined as 53 rather than 40 hours and overtime is calculated on 28-day rather than 7-day periods, the difference in fire department rate of pay versus that of departments with standard 40-hour workweeks is obvious. Fire officers can use this comparison in staffing and budget discussions with public administrators and other public officials.

A comparison of the compensation paid to fire fighters under the 40-hour-week provision versus compensation paid under the 53-hour-week/28-day work period demonstrates the impact of the 1974 FLSA amendments. The following tracks a 24/48 platoon shift pattern for three 28-day work periods. The fourth work period repeats the pattern of the first shift.

Standard hours for a 28-day (4-week) work period:
$4 \times 53 = 212$ hrs

Work Period I (beginning Sunday)

Week 1	72 hours	*Sun*	*Wed*	*Sat*	
Week 2	48 hours	*Tue*	*Fri*		Total Hours
Week 3	48 hours	*Mon*	*Thr*		240
Week 4	72 hours	*Sun*	*Wed*	*Sat*	

Overtime worked = 240 hours – 212 standard hours = 28 hours overtime

Work Period II

Week 1	48 hours	*Tue*	*Fri*		
Week 2	48 hours	*Mon*	*Thr*		Total Hours
Week 3	72 hours	*Sun*	*Wed*	*Sat*	216
Week 4	48 hours	*Tue*	*Fri*		

Overtime worked = 216 hours – 212 standard hours = 4 hours overtime

Work Period III

Week 1	48 hours	*Mon*	*Thr*		
Week 2	72 hours	*Sun*	*Wed*	*Sat*	Total Hours
Week 3	48 hours	*Tue*	*Fri*		216
Week 4	48 hours	*Mon*	*Thr*		

Overtime worked = 216 hours − 212 standard hours = 4 hours overtime

Total overtime = 28 + 4 + 4 = 36 hours for the three work periods.

For the entire year, the amount of overtime would be either 148 or 173 hours, depending on which work period schedule occurred first (or last).

In a 40-hour workweek, a 28-day pay period would have (4 × 40) 160 hours of standard time. Thus, the 40-hour workweek with the 24/48 shift would result in 80 hours of overtime in Work Period I and 56 hours of overtime in Work Periods II and III. The savings of the 53-hour week, 28-day work period provision is obvious.

Comparison of Overtime Hours under the FLSA 53-hour Week and a 40-hour Week

	Work Period I	*Work Period II*	*Work Period III*
Total hours worked	240	216	216
Overtime worked under FLSA 53-hour week	28	4	4
Overtime worked in a standard 40-hour week	80	56	56

A < B = lower costs to department

A number of municipalities began to experiment with alternative staffing patterns following the 1985 amendments. Some have proven successful, others have not. The most common alternative patterns consist of fourteen 10- and 14-hour tours of duty per week. Overtime can be eliminated through increased staffing. However, the cost of fringe benefits for the increased number of employees often negates any savings gained from reducing overtime.

Police departments have increased staffing during peak demand periods and reduced staffing during slow periods to minimize overtime and provide better service during peak demand. The use of the platoon or team approach to fire fighting makes this arrangement more difficult for the fire service because it is not simply a matter of adding individuals to the assigned time periods. Another challenge has been to identify the time periods of high demand. Further research, aided by technology, may indicate the solution: "As the fire service becomes more adept at computerized analysis of emergency incidents and develops a good base of 5 to 10 years of response data, it is very likely that a staffing arrangement similar to police staffing will be used in the future."[2]

Additionally, the requirements for overtime compensation have resulted in improvements to mutual-aid agreements. This is part of the trend toward regional fire protection: shared resources reduce costs.

Smaller fire departments and combination departments often take a different approach to scheduling. Volunteer or paid on-call fire fighters are used to supplement the full-time fire fighters, reducing the number of overtime hours worked. This works especially well when on-call personnel are assigned in platoons and have the opportunity to train and respond with the full-time fire fighters on a regular basis. This enables the on-call and full-time personnel to build a team atmosphere, facilitating effective response.

While career fire fighters may oppose reductions in overtime because they rely on the income to supplement their salaries, departments may not be able to afford the increased overtime costs to cover staffing requirements. Minimal staffing would make overtime available, but the community may not receive the optimal level of fire protection for its financial investment in the department. This violates the trust of the citizens. In addition, and of more immediate relevance to most fire fighters, the safety and effectiveness of fire fighters is decreased when staffing is minimal. The two most important issues in scheduling and cost management are providing optimal fire protection to the community and ensuring the safety of fire fighters during emergency response situa-

FIGURE 4-1 Today's fire fighters are highly trained professionals who keep current on all aspects of the job.

tions. If this can only be accomplished through the use of part-paid personnel, then that program must be adopted.

Given the increasing (and probably continuing) financial restrictions on local communities and, consequently, fire departments, innovative scheduling and use of paid on-call personnel will be required. Exactly when and how this should take place depends on the services required by the individual community. Each department should analyze its service demands and compare them with the use of resources available in order to ensure those resources are maximized. Techniques for accomplishing this are presented in Chapter 9, Needs Assessment. In addition, many of the planning techniques presented in Chapter 10 are helpful.

Occupational Safety and Health Administration (OSHA)

The Occupational Safety and Health Administration (OSHA) is charged with ensuring the safety of employees in the workplace. It mandates safe work practices and the use of personal protective equipment. The technical requirements of OSHA regulations are extremely detailed and will not be thoroughly discussed here. The purpose of this section is to acquaint fire officers with the major regulations that have an impact on administrative responsibilities. For broader discussion, we recommend consulting the more technical resources listed in the reference section at the end of this chapter.[3] For specific detail, consult the original OSHA regulations: *Code of Federal Regulations,* Title 29, Parts 1900–1910, Revised as of July 1, 1992.[4]

States must comply with federal OSHA standards unless existing state standards meet or exceed OSHA requirements[5] and the state has signed an agreement following the rules in Part 1901, Procedures for State Agreements. Under Part 1901, the state must enter into an agreement with the Department of Labor certifying that the state will enforce the standards specified in the agreement. If existing state standards do not meet OSHA requirements or if the state does not have an agreement with the Department of Labor, OSHA regulations prevail. A number of states have simply adopted the federal regulations.

States may exceed OSHA requirements with one exception. Under OSHA 1910.120, all states must comply with the *Hazardous Waste Operations and Emergency Response Standard* to ensure nationwide uniformity. Since compliance with this section of the OSHA regulations is universal and directly relevant to the fire service, a brief discussion follows.

OSHA Hazardous Waste Operations and Emergency Response Standard

The federal Superfund Amendment and Reauthorization Act (SARA) Title III, Emergency Response Planning and Community Right-to-Know, addresses the issue of hazardous materials (hazmat) emergency response in general terms, requiring only that some form of first responder training should take place at a local community level. The exact form of the training and the target audience for this training is not clearly defined by the standard. Thus, the level of training and preparedness will vary according to the local hazmat response needs and community response capability. Since these two factors vary significantly from community to community, or industry to industry, a strong hazmat training standard was needed. To address the problem, OSHA promulgated the *Hazardous Waste Operations and Emergency Response Standard* (OSHA 1910.120).

OSHA 1910.120 requires that those employees managing uncontrolled hazardous waste sites and those engaged in routine or intermediate handling and disposal activities receive training relating to the materials at the site. These training provisions specifically require that employees be provided with the knowledge and skills necessary to perform hazardous waste cleanup operations with minimal risk to their safety and health.

The OSHA standard also addresses training requirements for off-site emergency responders to hazmat incidents. Employers such as fire departments, emergency medical and first aid squads, and fire brigades are specifically identified as those local agencies that are covered by the standard. The requirements state that:

> *Training for handling emergency responses involving hazardous substances shall be conducted on a monthly basis and shall be at least 24 hours annually. The training shall include as a minimum recognition of hazards, selection, care and use of personal equipment and safe operating procedures to be used at the incident scene.*
>
> —OSHA 29 CFR, 1910.120(l)(3)(i)

The training includes: (1) proper use of a self-contained breathing apparatus (SCBA), (2) correct procedures for using personal protective equipment (PPE), (3) incident mitigation procedures, (4) the use of specialized hazmat equipment, (5) research techniques for determining the material involved in the release or spill, and (6) preplanning methods for hazardous materials incident operations. The OSHA standard further identifies the makeup and purpose of a hazardous materials response team and specifically requires advanced levels of training in areas such as the use of specialized chemical-protective clothing, techniques and procedures for stopping or controlling leaks, and decontamination.[6]

Hazmat response team members are also required to participate in a personal physical examination program. Fire fighters, law enforcement, and other first responder personnel are not required to comply with the physical examination portion of the standard unless they

FIGURE 4-2 Fire officers must ensure that their departments meet federal OSHA standards for responding to hazardous materials incidents.

have been exposed to a hazardous or toxic substance under emergency conditions. The intent of the physical examination for hazmat response team members is to provide medical professionals with baseline medical information needed to treat a hazmat responder victim in the event of an accidental exposure to a hazardous substance. It will also identify any previously undetected physical deficiency or chronic condition a prospective hazmat responder may have that would impede the performance of that person under emergency conditions. This requirement serves two purposes. It ensures that those individuals serving on a hazmat team are in good health. It also enables the hazmat team's medical advisor to monitor the health of the team members and provide quick medical treatment should exposure be identified following any incident.

The health and safety of individual emergency first responders, publicly or privately employed, are addressed in OSHA 1910.120. Fire department officers need to be aware of and responsible for complying with these standards to maintain the safety of all fire department personnel.

To remain current with changes to legislation pertinent to the fire service, fire officers must use resources such as professional associations and read the available journals and newsletters covering legal and professional responsibilities. (See Chapter 7, Interacting with Other Public Administrators, for more specific information.) Maintaining high levels of competency and professionalism reduce the potential for litigation and loss of life in a hazardous materials or other emergency.

Workers' Compensation

Workers' compensation provides payments to a fire fighter injured on the job or to survivors of a fire fighter killed in the line of duty. This coverage is being extended to volunteers and part-paid personnel in some states. The provisions of workers' compensation laws vary from state to state. Injuries covered by workers' compensation must "arise out of and in the course of"[7] employment. Some states identify the specific types of injuries that are covered and how they must occur. Other states have broad definitions subject to interpretation.

Almost all states require a waiting period, generally of 1 to 2 weeks, before compensation is available. Most states provide three types of benefits: death benefits, medical expenses, and wage replacement payments. Medical expenses are usually paid following the injury and do not include the waiting period. They cover "usual and customary" costs of medical care for the injury sustained. Death benefits are a one-time lump sum payment to the survivor, and wage replacement usually ranges from 50 to 70 percent of an average week's wages. Wage replacement begins after the state-mandated waiting period.

Workers' compensation is a form of insurance that employers pay into a state fund to cover claims by employees. Like insurance, the amount of payment depends to some extent on the number of claims against the employer. Thus, fire departments with a substantial number of workers' compensation claims pay higher rates. The cost of avoidable injuries continues beyond the injury and/or death of individual fire fighters.

Laws Governing Equal Compensation

Equal Pay Act

In 1963 Congress passed the Equal Pay Act as an amendment to the FLSA. It states:

> *No employer shall discriminate between employees on the basis of sex by paying wages to employees less than the rate at which he pays wages to employees of the opposite sex for equal work on jobs which require equal skill, effort, and responsibility, and similar working conditions.*
>
> —29 USC 201-77 Stat. 56

This legislation requires all fire fighters, regardless of gender, to receive the same monetary compensation and benefits package.

Civil Rights Act

The Civil Rights Act of 1964, as amended in 1972, prohibits unequal payment of wages on the basis of race, gender, color, religion, or national origin. It extends the Equal Pay Act to cover other minority groups in the workforce. The Civil Rights Act is enforced by the Equal Employment Opportunity Commission (EEOC), which can bring class-action suits against fire departments with discriminatory compensation systems.

Age Discrimination in Employment Act

The Age Discrimination in Employment Act of 1967, amended in 1978, protects employees and prospective employees between the ages of 40 and 70 from unfair treatment due to age. The employer cannot deny pay increases, bonuses, or benefits on the basis of age. The EEOC is also charged with enforcing this law.

Most fire departments do not violate the compensation requirements of these laws. Generally, compensation is the same for all fire fighters with the same classification. Sometimes there is a differential based on seniority, but seldom does one find compensation differences based on personal characteristics. Nevertheless, it is important for fire officers to be aware of these laws and to monitor their enforcement in the fire department.

Equal Employment Opportunity and Affirmative Action

The issue of unequal treatment of members of minority groups extends beyond the compensation question and is the area where fire departments are more likely to experience difficulties. Although Equal Employment Opportunity (EEO) and Affirmative Action (AA) are often discussed together, they are, in fact, based on different laws and have different intents. Simply stated, EEO provides for equal treatment and AA provides for preferential treatment of certain groups on the basis of social inequity. The enforcement of both EEO and AA is the responsibility of the Equal Employment Opportunity Commission or can be delegated to the state.

EEO is mandated by Title VII of the 1964 Civil Rights Act and AA is based on the 1972 Equal Employment Opportunity Act. Title VII prohibits fire departments from making employee or applicant personnel decisions based on race, color, religion, gender, or national origin. AA not only prohibits discrimination, but requires fire departments to take positive steps to reduce underrepresentation of minority groups in the workforce through the preparation and implementation of affirmative action plans. The EEO applies to all fire departments. AA, however, may be based on either voluntary or involuntary compliance.

Voluntary AA compliance results from a public agency, department, or governmental entity preparing an affirmative action plan that identifies underutilization of a specific part of the community workforce, establishes full utilization of this group as a goal, develops a plan to achieve full utilization, and makes progress toward implementing the plan. Involuntary AA compliance is the result of an investigation based on a formal complaint from an employee or applicant that results in (1) a negotiated settlement whereby the employer agrees to modify employment and hiring practices, (2) a consent decree (an agreement between the employer and the compliance agency) approved by the court, establishing an affirmative action plan that will be monitored, but does not require the employer to admit guilt, or (3) a court order based on findings of a court that the employer is in violation, must make restitution, and must rectify employment and hiring practices, which will be monitored by the state enforcement agency and, in some instances, the federal court system.

Government entities in many urban areas have affirmative action plans in place. Fire officers, especially those in charge of hiring, training, employee review, and promotions, should become familiar with these plans and identify how they affect the fire department. The level of management that should be familiar with AA plans ranges from the fire chief in charge of maintaining AA in the department to the line officer lieutenant responsible for daily monitoring of activities and personnel review.

Equal employment opportunity requires broad-based understanding of its intent throughout the department. Hirings and promotions are based on the top score on the written test and the best performance in the physical fitness and agility test. Affirmative action requires that when members of the underrepresented class meet the minimum standards (or in some cases are in the top three or five), they receive the offer of employment or promotion. This is done in an attempt to address previous wrongs to a given group and to achieve social equity by allowing each group the opportunity to gain employment.

The only exception to the EEO requirements is provided in Section 703(e) of the Civil Rights Act, which states that religion, sex, or national origin can be considered if such an attribute is a "bona fide occupational qualification (BFOQ) reasonably necessary to the normal operation of that particular business or enterprise." Race is never a legitimate BFOQ, but gender, age, religion, and national origin have been substantiated as valid in particular instances. Some departments in the fire service attempted to utilize the BFOQ to exclude women on the grounds that women were not suited for strenuous jobs and that tradition mandated that the fire service employ males exclusively. However, the courts ruled that tradition was not grounds for discrimination and that women who could meet the physical requirements of the job—which were reasonable and representative of normal response expectations—and who meet other non-gender-biased criteria, must be allowed both employment and equal opportunity for career advancement.

Fire departments and officers that have refused to allow women to

FIGURE 4-3 Women are guaranteed equal employment opportunities and have proven to be highly qualified fire fighters.

respond to fires, wear turnout gear, or train for all aspects of fire control and mitigation are violating EEO mandates and have been found liable when complaints have reached the investigation stage or court system. Fire officers who refuse to allow any employee the opportunity to develop his or her abilities and advance his or her career are guilty of violating EEO and place both themselves and their department in jeopardy of a discrimination lawsuit.

Despite the fact that over 51 percent of the population and approximately 40 percent of the workforce is made up of women, women are a minority in the fire service. Women are joining the fire service in growing numbers; however, they often face discriminatory and demeaning attitudes and treatment that limit their career development. Additionally, women are not always allowed to use their talents and knowledge to benefit the fire service. Instead, they are relegated to a few jobs and are not allowed to participate in all aspects of the department. This is not only a waste of talent, but in many cases illegal.

Sexual stereotyping causes people to dislike behaviors in women that are favored in men. A study has shown that women leaders receive negative (verbal and nonverbal) feedback when they display the same behaviors that are positively reinforced in men.[8] This type of sexual discrimination can be direct or indirect.

Indirect sexual discrimination can be seen in condescending attitudes, special training, and assignments given to women fire fighters. For example, assigning the female fire fighter on an engine company to running the pump only as restricting participation in attacking the flames, especially when the pattern of assigning one individual exclusively to the pump is not followed on other engines, implies that the female fire fighter is weaker or less courageous. Limiting hazmat training to men or refusing to allow women to drive apparatus are other examples of sex discrimination. This implies that the physical differences between male and female fire fighters make women incapable of performing all fire fighting duties. However, if male and female fire fighters meet the same entry and continuing qualifications, there is no justification for this bias. Women are discriminated against as a group in the same way that ethnic and racial cultures are. As a result, when dealing with diversity, fire departments need to include women as well.

Discriminatory hiring practices aimed at racial, ethnic, and religious minority groups are also illegal. All individuals who meet the requirements of the position and qualify through an unbiased testing procedure must be allowed the same opportunity to secure the job of fire fighter. When setting the requirements for the position, it is necessary to ensure that all of the requirements relate to job performance. Height and weight requirements have been struck down as unreasonable since a person's height has nothing to do with the ability to perform the functions of a fire fighter. Height had been a requirement in many departments because of the need to access various compartments on an apparatus. The courts ruled, however, that the apparatus could be modified without mitigating the individual's quality of performance and, therefore, the height argument was invalid. A better test requirement is one of physical stamina or ability. For example, a person could be required to ascend a certain number of stairs in full turnout gear, including SCBA, within a reasonable period of time. This ability relates directly to the requirements of the job and is neutral to gender and race.

Most of the EEO complaints against fire departments could be eradicated simply by instituting and following fair guidelines for hiring and promoting individuals. Fire departments encounter problems when personal biases interfere with evaluating the quality of an individual's job performance or potential.

Sexual Harassment

Sexual harassment violates Title VII as one version of sex discrimination. The EEOC adopted guidelines interpreting sexual harassment in 1980. Verbal and physical conduct of a sexual nature is harassment when any one of the following conditions is met:

- Submission to such conduct is either explicitly or implicitly made a term or condition of an individual's employment
- Submission to or rejection of such conduct by an individual is used as the basis for employment decisions affecting such individuals
- Such conduct has the purpose or effect of substantially interfering with an individual's work performance or creating an intimidating, hostile, or offensive working environment

This last condition is extremely important to understand. For example, if sexual jokes in bad taste create an offensive environment for male or female fire fighters, they are being subjected to sexual harassment. In these cases, not only are the fire fighters telling the jokes at fault, but so is the immediate supervisor who allows the joke-telling to take place, and the senior officers, who are responsible for enforcement of sexual harassment policies. Even the fire department as an entity can be named as party in a discrimination suit.

> *If sexual harassment occurs, "an employer is responsible for the acts of its supervisory employees or agents, regardless of whether the acts were authorized or forbidden by the employer and regardless of whether the employer knew or should have known of the acts."*
>
> —*EEOC Guidelines,* 1980

Fire officers cannot ignore cases of sexual harassment. Fire fighters must act like the professionals they are, and it is the duty and responsibility of fire officers to see that they do so. Ignoring a situation of harassment (sexual, racial, ethnic, or religious) is tantamount to condoning the act and implies complicity. It is for this reason that the law extends the liability to supervisors and employers. Employers and officers may mitigate their liability when charged with allowing sexual harassment to occur if "immediate and appropriate corrective action is taken." Fire officers cannot simply let a situation pass; they must take immediate and decisive corrective action. This sends a clear signal as to what conduct is appropriate and what is not.[9]

Sexual harassment is much more than requesting or requiring sex in exchange for employment, assignment, or promotion. It is anything of a sexual nature that interferes with an individual's ability to perform a job or advance his or her career. The test of whether a statement or action is sexual harassment *lies in its impact on the employee,* not in the intent of the perpetrator! Thus, if lewd statements are upsetting to the point where they interfere with an individual's job performance, they are a form of sexual harassment.

For example, the first female fire fighter hired by a large metropolitan fire department was assigned to an engine company. Her assigned duties were the same as those of other fire fighters assigned to the engine, but her position on the engine put her closest to the pressurized water extinguisher that was mounted near her assigned jump seat. The captain instructed her that anytime the engine responded to a structure fire, she was to secure the pressurized water extinguisher immediately and carry it to the front of the fire.

Shortly after joining the engine company, they responded to a large industrial warehouse fire. Per her orders, she left the engine with the pressurized water extinguisher in hand even though the building was fully involved. The assistant chief in charge observed this action and yelled, "What are you going to do, honey? Put the fire out with that?"

Following the incident the female fire fighter filed a grievance and the assistant chief was severely reprimanded. The comment cast doubts on her ability as a fire fighter and on her judgment of the situation, even though she was carrying out her company officer's direct orders.

Because the fire service has traditionally employed males exclusively, mitigating sexual harassment poses a challenge to the fire service. Including females in the ranks of departments has stirred some resentment and unfair treatment that diminish a female fire fighter's ability to perform her job and, consequently, her opportunity for career advancement. Behaviors that were acceptable in an all-male environment may now cross the boundaries of acceptable behavior and lead to charges of sexual harassment. Many fire departments have played jokes on rookies as an initiation rite.

Behavior like this must be eliminated. It is neither appropriate nor legal to permit harassment to continue. Fire officers must make a deliberate effort to create and maintain standards of equal treatment for all personnel.

Fire departments can take the following measures to prevent sexual harassment:

- Develop a clear and comprehensive policy against sexual harassment
- Develop and institute appropriate sanctions against such behavior—*and enforce them*
- Actively inform employees of their rights
- Actively communicate the policy to all employees
- Identify and communicate the grievance procedure
- Encourage employees to file a grievance when having cause
- Sensitize all fire officers and fire fighters to the moral and social offensiveness of sexual harassment

Although it is not an acceptable excuse, it is true that some fire fighters—and officers—may behave in an offensive manner without realizing that they are insulting someone. Sensitivity training is an effective tool against sexual harassment and racial discrimination because it broadens awareness of these issues.

Fire officers must take a proactive approach to eradicating sexual and racial harassment and discrimination in the fire service. These practices decrease the effectiveness of the service and create animosity among its members. Respecting each fire fighter as an individual, offering the same opportunities to everyone, and hiring and promoting on merit and performance will alleviate the residue of old biases. Decisions made on the basis of "old boy" networks, racial or sexual biases, or petty grudges have no place in the fire service of today or tomorrow.

THE MULTICULTURAL FIRE SERVICE

Fire departments recruit members from their communities. Consequently, as the demographics of a community change, the demographics of the fire department change as well. The court rulings and EEO mandates discussed in the last section were enacted to ensure that this takes place. Sometimes this change is slow, but it will, and must, take place to provide service that meets the unique demands of each community and offer careers to all qualified members of those communities. More cultural diversity is found on urban fire departments, as the composition of these communities is reflected in the training classes of new recruits. This section of the chapter offers guidelines for managing these changes in a positive way. (Questions of recruitment and training are covered in Chapter 5, Hiring Practices.)

Increased diversity of people in an organization creates a need for a different organizational culture to fully utilize the contributions of all members of the organization. This is the cornerstone of the culture of diversity philosophy. For the purpose of this discussion, **cultural diversity** refers to the existence of an assortment of cultures in both the community and the fire service. The **culture of diversity** refers to an organizational culture that values the differences among the members of the organization.

The discussion of the cultural diversity in the U.S. in Chapter 3 provides a background for understanding why successful organizations are changing. Organizations that learn to utilize the particular strengths of each employee rather than forcing conformity to the existing culture of the organization are better able to meet the needs of changing communities and to provide services.

To establish a culture of diversity is to focus on the quality of the work environment and improved utilization of skills of all employees. It moves beyond the requirements of affirmative action. Affirmative action and equal employment opportunity are programs initiated and mandated by the government to address the *quantity* of minority representation in the organization. An organization that values diversity and works toward creating a culture of diversity voluntarily balances minority representation and redesigns the quality of operations to utilize the special contributions of the diverse cultures in the organization. EEO is driven by legal mandates. Valuing cultural diversity is driven by improved productivity and enhanced service to the community. EEO is designed to equalize representation of minorities in the organization; valuing cultural diversity takes the opportunity provided by diversity to enhance the organization's operations.

The traditional attitude toward cultural diversity in the U.S. focused on creating a "melting pot," wherein minority cultures would assimilate into the majority culture. Today's view of cultural diversity appreciates the differences among various cultures and seeks to promote the integration of these differences into the culture at large, or at least allow differences to exist in harmony with one another. Affirmative Action and EEO programs were developed as a reaction to institutionalized bias. Valuing cultural diversity is positive and proactive. It takes advantage of opportunities presented by diversity. This discussion of cultural diversity looks at going beyond the legal requirements of representation in the fire department to how representation of each culture in the community can enhance the performance of the fire service.

Achieving Positive Cultural Diversity

A clear message needs to be sent to members of a fire department that values diversity among its members. One way to communicate this message is to include it in the fire department's goal or mission statement. The employees at NYNEX Mobile Communications Company developed the following statement to reflect the organization's goals:

> *Because we must reflect the diverse society we serve, it is critical to our business that we become more diverse. Thus, as an organization, we are committed to creating a culture that promotes mutual respect, acceptance, cooperation, and productivity among people who are diverse in work background, education, age, gender, race, ethnic origin, physical abilities, religious beliefs, sexual/affectional orientation, and other perceived differences. Understanding and valuing differences will maximize the growth and development of our employees and meet the needs of our increasingly diverse customer base.*[10]

A goal statement at Stanford University, written by a group of students, faculty, and staff, reads:

> *Having achieved a strong measure of racial diversity. . . we must now make a new commitment. We must make the transition from numerical diversity to interactive pluralism. We must define a new vision for the University as a community where all ethnic groups can engage their differences in a process of mutual enrichment. . . . We believe that gender, racial, ethnic, cultural, religious, and other individual or group differences enrich the educational and social environment where we teach, learn, live, and work. These differences, rather than inhibiting communication and concord, present us with singular opportunities find mutual understanding and respect.* [11]

When developing a mission statement for a fire department, follow the rules outlined in Chapter 6, Planning. In addition, be sure that diversity is discussed in broad multicultural terms that clearly define the rules of interactive behavior and include all current and potential future members of the department. The final critical element to stating and promoting the value of diversity up front is identifying the benefits to the fire service and/or department. This ensures that the members of the department understand the importance and value of diversity among their fellow members.

Obviously, this commitment to diversity must start with the leaders in the organization—especially those at the top. **Pluralistic leaders,** as discussed in Chapter 2, value and support cultural diversity. Leaders throughout the organization can support multicultural contributions through both formal and informal recognition. This is key to the true integration of cultural diversity as a positive aspect of the fire service.

Leaders at the top of the organizational hierarchy must take a visible stand on valuing diversity in the department and in the fire service as a whole. The contributions of minorities and women need to be recognized and shared. In this way members of the fire service will learn how to recognize alternative contributions that may be made by different cultures and groups of people.

The use of **diversified work groups** is one way to benefit from cultural diversity in an organization. The fire service should encourage if not require the participation of minority group members in decision making. This enhances end decisions by providing a broader view of the problem along with diverse, alternative solutions. This also demonstrates that management is sensitive to exclusion and wants to include different cultural viewpoints in the operation of the department.

Bilingual fire fighters can offer specialized language classes to paramedics and dispatchers. Representatives of ethnic groups can be asked to explain customs. Public education programs staffed by diverse fire fighters reflect diverse communities and result in more affinity with the audience. These are examples of taking advantages of the opportunities provided by diversity.

Departments with few minority employees should avoid overburdening minority individuals with assignments to numerous work groups simply because they are the sole representatives of their culture, however. In these cases, it may be necessary to prioritize issues important to those minority members.

The fire service can also begin to **reward behavior that values diversity.** A shift captain who integrates female and Hispanic members of the department by assigning them roles in training incidents should be recognized by his or her superior. Positive actions should be recognized and rewarded. Strained financial resources and restrictions of labor agreements may limit that reward to public recognition, but even if the reward is limited, acknowledging the behavior sends the message that using and promoting diversity are supported by the department.

Conversely, negative behaviors should be reprimanded. For example, in one full-time department, it was the practice of a battalion chief to tell a joke just after shift change. This was supposed to "cheer up the troops" for the day. However, many of these jokes contained ethnic slurs and others were derogatory toward women. By not censoring this chief, the fire department administration was tacitly endorsing sexism and racism. Consequently, the department experienced a high level of tension that resulted in numerous discrimination grievances and two major discrimination suits. Remember, *silence connotes approval. Prejudice cannot be approved of or ignored.*

Informal coaching and tutoring; sessions or programs enable the majority members of the department to share operational procedures with minority members. In addition, the minority members have the opportunity to tutor majority members in the cultural differences of minority groups in the community. This enables both groups to share knowledge and to have a positive impact on the organization. These programs provide a forum for sharing information and ideas, with each "partner" contributing rather than one giving and the other receiving. In a culturally diverse fire department, this approach can be extended to mentoring programs by enabling the protégés to share and contribute as well as learn. Mentoring is primarily a one-on-one situation, whereas coaching and tutoring can occur in group settings.

POSITIVE COMMUNICATION

Positive communication is a vital component of all successful aspects of human resources management. Although much of this discussion relates directly to the relationship of communication to culture, it can be applied to all aspects of daily activities. Encouragement and positive communication (in giving orders or informal discussions) have a much more successful impact on the morale, motivation, and consequent performance of fire fighters. As you read the following discussion, relate the techniques to all aspects of the department, whether diversity is involved or not.

Too often language reflects the biases of our culture. It incorporates stereotyping and demeaning terms about others. In a multicultural workforce and fire department, it is imperative that everyone consciously use appropriate language. Appropriate language is positive and inclusive. It recognizes the value of diversity and the contributions of others, and avoids demeaning terms.

Not only can words be demeaning, but the direction and tone of the communication used can be insulting. For example, a captain who does not address an African-American fire fighter directly is clearly building a distance and sending an unspoken message that this fire fighter is neither respected nor valued. Using a condescending tone of voice when discussing complaints or questions from female members of the department is also a form of negative communication and should not be tolerated. Often, these types of communication happen unconsciously, and the speaker may be genuinely unaware of the offensive nature of his or her delivery or tone. Becoming aware of more subtle forms of racism and sexism requires education, self-awareness, and willingness to listen to others and change one's behavior when a problem is revealed.

Choice of words is a frequent source of perpetuating stereotypes and devaluing diverse cultures. Calling female fire fighters or dispatchers "girls" is demeaning and sets them apart from others in the department. Phrases such as "you people," "you folks," and even "they" serve as a means to separate and stereotype, indicating differences and barriers rather than commonalties. Any stereotypic generality degrades a particular group of people. As mentioned before, humor that relies on putting someone down, especially members of a specific group, is inappropriate and in poor taste.

Valuing diversity requires members of the fire department to appreciate each individual and his or her heritage and contributions. There is no place for a negative and demeaning comment, slur, look, sneer, or attitude. These negative communication practices simply serve to rob the fire service of valuable assets. Resources are scarce and will become even more so. *No* department can afford to lose the contributions of any of its members. Fire officers of the future must communicate effectively with a variety of groups in their department as well as in the community. In addition, they must ensure that others in their department communicate in a positive, supportive, and effective manner.

DEALING WITH PERSONAL PREJUDICES

Studies have shown that many people who display negative, nonverbal reactions to members of minority groups in an organization seem to be unaware that they are doing so.[12] Many people have been taught to view differences with suspicion. We don't like what we don't understand. In addition, people often resist change. Change results in uncertainty and requires effort and growth. This takes work.

Fire service officers must serve as examples to the department. Loden and Rosener offer the following five-step approach to dealing with personal prejudices and stereotypes. This approach does not focus on discovering the reasons for behavior. Instead, it concentrates on counteracting negative behavior; it is a change-oriented model. Fire officers can follow the model and recommend it to other members of the department who may benefit.

1. *Accept responsibility for the problem.* Denial is one of the leading causes of discrimination. Officers who deny that any of their actions are discriminatory or prejudicial prevent themselves from growing and developing. They signal others in the department that prejudicial activities are acceptable. Fire officers who view monitoring discriminatory behavior as the responsibility of the EEO officer are denying their responsibility for increasing an appreciation of cultural diversity. The responsibility for minimizing negative behavior based on personal prejudice belongs to every member and every officer of the department. Accepting responsibility does not mean piling on guilt; it means taking a step to solve the problem.
2. *Identify problem behaviors.* Prejudicial behaviors must be openly identified and noted. Individuals who exhibit these behaviors must be made aware of their actions. Often becoming aware of the negative nature of these actions is enough to cause an individual to change. Fire officers must identify negative behaviors in themselves and others. *Prejudice or prejudicial behavior* means to "pre judge": to make judgments without facts or information; to base activities on assumptions. Ignorant or untrained actions are dangerous in the fire department as well as on the fireground.
3. *Assess the impact of behavior on others.* Once negative behavior has been identified, the fire officer must minimize the damage done. The effects of prejudicial behavior must be identified and corrected. The ramifications of such behavior must be clearly understood by personnel. This aids in understanding why such behavior cannot be tolerated in a multicultural or culturally diverse organization.

4. *Modify negative behavior.* The next step is to correct improper behaviors, both in one's own and other's behaviors. Anyone identified as participating in stereotyping behaviors must be taught to change how they act. Sometimes those who are affected by the negative actions can assist in correcting the behavior: they can identify acceptable alternatives.
5. *Obtain feedback.* Modified behavior may not be an improvement over the initial behavior. It is imperative that fire officers get feedback on the changes from the affected minority group. Open communication with all elements of the department is a requirement for enhancing a culture of diversity. Once the prejudicial behavior is changed, both the individual and other fire officers need to know that the new behavior is acceptable. This enables both the individual and the department to continue to grow by defining the limits of unacceptable behavior and providing a formula to create change.

This five-step approach requires constant repetition. An individual with one negative behavior probably practices others. Continuing the five-step evaluation of behavior provides a prescription for continued growth and change. As leaders, fire officers must take the first step in knocking down barriers to different cultures in the department. By recognizing prejudicial behaviors and providing an objective method to cultivate positive behavioral changes, fire officers can set the tone for the rest of the department.

LABOR AGREEMENTS

Labor agreements are negotiated agreements between fire fighters and management, usually senior officers, that set forth the terms of employment. The labor agreement may cover many issues and be very detailed, or it may cover only a few areas such as wage compensation, hours, and benefits. There are seven areas of employment that are usually addressed by labor agreements: (1) compensation and benefits, (2) working conditions, (3) job security, (4) discipline procedures and individual rights,

(5) union security, (6) management prerogative, and (7) contract duration.[13] The exact topics considered in bargaining are dependent on state legislation. Some states mandate certain topics as part of any public-sector bargaining process. Other states limit what can be negotiated. Each fire officer should be aware of the specific state laws applying to his or her department. Once the labor agreements are set during the negotiations, any changes must be agreed to by both parties. This requires fire officers and fire fighter union representatives to sit down formally and discuss changes. If an agreement can be reached, the contract is modified. Although this sounds tedious, it can be more of an administrative formality if there is good communication and a good working relationship among the fire officers and fire fighters.

The International Association of Fire Fighters (IAFF) is the primary union representing full-time, career fire fighters. The IAFF provides a variety of services to its membership, including:

- Technical assistance with OSHA requirements
- Labor relations expertise utilizing in-house personnel and field representatives
- An extensive educational seminar program
- Legislative representation at the national level
- Information on testing and performance of various fire fighting equipment
- Public relations guidance

Organized in 1918, the IAFF is affiliated with the AFL-CIO. Any person who is engaged as a permanent and paid employee of a fire department is eligible for active membership through the chartered locals, state or provincial associations, and joint councils. The Association's official publication, *The International Fire Fighter,* is distributed monthly to its 175,000 members. Other publications include the annual "Death and Injury Survey," a "Public Relations Manual," and "Salary and Working Conditions Surveys."

Both fire fighters and officers can benefit from some of these services. Fire fighters need to learn to use IAFF resources when in situations other than "attack mode," during negotiations. Likewise, a positive relationship with the union, emphasizing mutual goals and cooperation, benefits the department as a whole.

Individual fire department IAFF unions have local autonomy. They can agree with national positions on issues or not. They can use information and assistance from the national office, or not. It is up to the local to decide. In addition, the local IAFF union can set standards for conditions of employment that differ from those recommended at the national level.

Grievances

The union contract is a set of rules agreed upon by both management and fire fighters. Both parties need to know what those rules are. If there is a disagreement in operations, check the union labor contract. Sometimes it is not so easy to settle a dispute, however, and the dispute results in a grievance procedure.

When a union files an unfair labor practice grievance against the management of a fire department, both parties must first sit down, discuss, and try to resolve the matter. If they cannot reach a resolution, they may have to enter formal negotiations and then arbitration. The exact procedures for each of these situations again varies by state as well as by local community. Some communities offer conflict resolution and mediation series to departments in labor disputes as a means of reaching a mutually agreeable solution while avoiding time-consuming and costly proceedings at the state level. Fire officers should know the procedures established by their community and state.

IAFF Vice President Jerry Holland points out that most grievances arise when either side—management or fire fighters—attempts to change something that has been previously agreed upon. Anything can be changed if both parties mutually agree, but this requires sitting down and communicating. Communication, or rather improper or lack of communication, is the source of most contract grievances. Holland identifies the position of most fire fighters in negotiation as: "If the answer is no, tell me why. Fire fighters want reasons instead of excuses."[14] The fire service is a paramilitary organization in which fire fighters are often called upon to think for themselves in highly stressful situations. They are skilled at looking at various options and solving problems. Thus, when involved in labor negotiations, the answer "because" as the reason for holding a position is not acceptable. It is for this reason that open, honest, and complete communication in labor negotiations and grievances works best.

This book has stressed that the fire officer must understand the community and the government entities in the area. Here is another area where such knowledge proves beneficial. Telling fire fighters that they cannot have raises because the city doesn't have money is probably not going to satisfy union negotiators. Showing them the realities of a declining tax base and falling revenue, however, may have an impact.

RETIREMENT PROGRAMS

Specific conditions of retirement programs vary by state as well as by community. Some states set uniform standards for all full-time fire fighters, while others set minimum standards that communities must meet. Some states also offer retirement program systems for volunteer or paid on-call fire fighters, which communities can elect to join. More and more states are enacting pension programs for volunteer fire fighters, due in part to the extension of workers' compensation coverage to volunteer and paid on-call fire fighters. The same reasoning that justified the extension of the workers' compensation benefits is used to justify the provision of a retirement pension program. These systems are typically coordinated at the state level, and communities rather than individual fire fighters have the option of participating. Some communities have their own retirement systems and, in states where this is permitted, the fire department personnel may simply be a part of that retirement system.

The amount of pension is usually determined by longevity (number of years served) and pay. The standard method of computing the base pay for pension is either the last year's salary or the average of the 3 highest years of pay. Thus, an assistant chief who was injured and voluntarily decided to move into the training division, even though it meant a reduction in rank and pay, and who then retired after 3 years, would receive a pension based on the highest 3 years' salary, earned when serving as assistant chief. The percent of base salary paid in pension is generally tied to the number of years of service in the department.

A cost-of-living adjustment (COLA) is usually part of a pension package. Just as COLA adjustments impact the annual pay, they also impact the amount paid in pension. Some public pensions have a COLA clause stipulating that whenever members of the department receive a COLA, former members still drawing pensions also receive the same percent COLA. Other agreements limit the percentage. For example, if the fire department agrees to a 4 percent COLA increase for all members and retirees' pension COLA is stipulated to be 75 percent of the COLA for active fire fighters, all those drawing a pension from the department would receive a 3 percent COLA in their pension payments.

Pension payments are becoming a source of increasing concern to public administrators. The cost is rising dramatically and the deficit or unfunded liability of several public pension plans is causing distress. Recent problems with private pensions have brought attention to this deficit, and many organizations and individuals are trying to identify solutions. Fire and police pensions are among the most expensive public personnel pension plans. This is due in part to earlier retirements necessitated by the strenuous physical demands and mental stress of these occupations.

The cost of employing a fire fighter is comprised not only of the salary and benefits, but also the cost of future pensions. This demand is almost doubling the cost of providing fire protection. For example, by 1980 the City of San Francisco was paying $0.90 toward retirement for every $1.00 paid to a fire fighter in salary.[15] Another way to understand the impact is to look at actuarial assumptions concerning the cost of a pension per fire fighter. For example, if data on a city show that the average fire fighter retires after 21.8 years of service and draws retirement pay for an average of 32 years, it will cost the city pension plan between $235,000 and $786,250 every time a fire fighter retires, depending on rank.

Some pension plans also include a health care package of some type. This provision is of increasing demand by fire fighters, as well as other employees. However, it is also an area receiving a great deal of resistance from administration because of unpredictably escalating costs. The cause of financial troubles for a number of pension plans has been the dramatic increase in health care costs and the associated health benefits in the pension plan. This is certain to be the focus of numerous discussions and a point of contention during labor negotiations.

SUBSTANCE ABUSE AND EMPLOYEE ASSISTANCE PROGRAMS

Abuse of drugs and alcohol is particularly dangerous in the fire service. Using these substances impairs one's own safety and endangers the lives of fellow fire fighters and the people in the community who depend on the fire service for protection and assistance.

Employee Assistance Programs are designed to provide counseling and treatment to substance abusers in order to rehabilitate them and reinstate them as responsible, contributing members of the fire department. These programs can be either voluntary or mandatory. Some of the voluntary programs are based on the employee's right to choose to enter the program or leave the department upon being presented with proof of abuse. Other voluntary programs allow fire fighters to enter the program at their discretion, when they feel that their addiction is beyond control and is interfering with their work and their life. These programs usually ensure confidentiality and often provide for temporary assignment of individuals in the program to areas of the department that are not high risk. Upon completion of the program—rehabilitation—the fire fighter is returned to active duty.

Employee assistance programs are cost-effective ways to save trained, experienced employees. It costs more to force retirement or to fire an employee, hire a new one, and train the new employee than it does to rehabilitate a current fire fighter. From a humanitarian and motivation standpoint, these programs are also highly beneficial for the department. Demonstrating that the department cares about its fire fighters and that it will give them a chance to turn themselves around builds loyalty. Even employees who don't use the service may be motivated by the department's concern.

These programs are usually offered through a contract with an existing social service agency or department in the community. A client referral system is set up, and the fire fighter is referred to professionals whenever the need arises. In communities where entry can be voluntary on the part of the employee without requiring referral, fire fighters can go directly to the agency and request assistance. When this occurs, the fire fighter and the counselor jointly make a decision as to whether the fire fighter's substance abuse is placing anyone in danger on the job; if so, the agency assists in arranging a temporary assignment for the fire fighter. The reason for the assignment remains confidential.

The topic of drug and alcohol testing is hotly contested. One side of the argument stresses that the nature of a fire fighter's job requires that there be no abuse, and that this must be ensured by public administrators. Accidents involving employees in public transportation have resulted in the establishment of precedents for mandatory drug screening in some public agencies.

Some unions have agreed to drug testing as a condition of employment, others have agreed to an annual screening, and others refuse to participate. One point of resistance to annual physical examinations is that some departments have used these as a means to conduct drug and alcohol tests without identifying them as such. This is inappropriate. If drug screening is to take place, it must be an open requirement of the department. If the department can show just cause, it will be able to secure this requirement. It is a good idea, however, to couple the requirement with an established employee assistance program in order to clearly demonstrate that the purpose of the program is to increase employee and community safety and offer rehabilitation to affected fire fighters. Hiding the testing process only builds suspicion and animosity.

EMPLOYEE DEVELOPMENT PROGRAMS

Chapter 6, Motivation and Discipline, shows that employee development is one of the greatest sources of motivation available. The ability to expand one's knowledge and improve technical as well as interpersonal skills builds confidence and keeps a job fresh and exciting; it prevents the job from becoming a boring routine.

Skills development may take the form of technical training in new methods of operation or in the use of new equipment. These programs offer fire fighters the opportunity to learn different aspects of fire suppression, which may be beneficial when someone who ordinarily performs those tasks is ill or injured and not available at the scene of an incident. Fire fighters can also develop areas of specialization, such as search and rescue or underwater rescue. This not only builds pride, but it also provides the department with motivated and capable specialists who can prove invaluable during emergencies.

Another area of development is that of human relations. This is especially important for fire fighters who aspire to become officers, as well as current fire officers. Working with people is the most important component of an officer's job. A fire officer must be able to motivate and communicate with others. This ability needs to be constantly refined and improved over time. Thus, employee development is an ongoing process in which all fire officers and fire fighters should participate.

Employee development for fire fighters can also incorporate human relations to assist fire fighters in getting along not only with their superiors, but also with the other fire fighters on shift, with members of other departments (such as police, road, and water departments), and with members of the community. Each of these groups of individuals requires a somewhat different approach to interaction. Fire fighters need the opportunity to learn the skills necessary to do this well.

Sensitivity training designed to assist an individual in understanding and dealing with differences should be a primary component of this employee development program. As communities change (see Chapter 3, Cultural Differences in the Community), fire departments need to educate their members about those changes and make them aware of differences in customs that they may encounter. Likewise, sensitivity and awareness training can mitigate many of the problems associated with the increased number of women and minorities entering the fire service.

Officer and fire fighter development programs can provide tremendous benefits to a fire department at relatively low cost. These programs make the department an exciting, dynamic place in which people enjoy working. This builds effectiveness and efficiency of operations. Everyone benefits—especially the community.

SUMMARY

This chapter has dealt with varied components of human resources management, ranging from legal requirements and labor agreements to fire officer development programs. Fire officers must understand not only the letter of the laws, but the intent of the laws. Generally, labor laws are enacted to rectify unjust treatment of employees. By being proactive in dealing with members of the fire department and addressing problems when they are small, fire service officers not only can make the department operate better, but may also be able to make further laws and regulations unnecessary.

CASE STUDIES

Case #1

Fire fighter/paramedic Michael Lewis, an African-American, requests a meeting with Chief Jeffrey and asks to be transferred to a different squad. When asked why, Lewis replies that he feels discriminated against by Captain Panner and he is worried that Panner's interference with the performance of his duties will eventually harm either a patient or another fire fighter.

When asked for details, fire fighter Lewis reports the following incidents. When responding to a motor vehicle accident, the squad encountered an automobile that had hit a tree. During extrication, Lewis concentrated on stabilizing a patient who was bleeding profusely from two head wounds and had at least two fractured ribs, a fractured femur, and unknown internal injuries. Paramedic Lewis called for Life Flight, believing that the patient needed a full trauma center. Captain Panner, a basic EMT and officer in charge of the scene, countermanded his order and told the EMT not to call, saying that they would transport by ambulance to the closest hospital (which was not a full trauma facility) more quickly. Lewis instructed the other EMT to take over monitoring the patient and called for Life Flight himself before returning to the patient. After the incident, Captain Panner berated Lewis in public using derogatory language.

During a garage fire, Lewis was the pump operator. He was supplying two 1½-inch lines, which were staffed by four fire fighters who had entered the structure. The engine, carrying 750 gallons of water, had not yet been connected to the hydrant by the second due engine. Captain Panner elbowed Lewis aside and opened the valve to the deluge gun. Lewis jumped in front of Panner and turned it off, fearing that the fire fighters in the structure might run out of water. As Captain Panner again reached for the deluge gun valve, the assistant chief arrived and took command of the scene.

1. What should Chief Jeffrey do?
2. There were witnesses at both incidents. Should they be involved?
3. Is this a case of discrimination? How can you tell?
4. Were Lewis's actions appropriate? What more might he have done? During the incidents? After each incident?

Case #2

Although a fire department employs a number of both minority and women fire fighter/paramedics, one of the assistant chiefs has noted that more promotions of women come from one engine company and fewer women resign from that company. Upon further investigation, he discovers that the company officer has instituted a training program in which each member of the company takes responsibility for each position on the truck in simulation exercises. Following the exercise, there is a thorough debriefing. As a result, the members realize that everyone, including the women on the shift, has the ability to do the job necessary during an incident. This professional respect and confidence has led to personal regard for the company members.

1. What should the assistant chief do?
2. Why did this training prove successful?

Case #3

Assistant Chief Woods has been on the department for 25 years and is currently in charge of fire operations. The department now has three female fire fighter/EMTs on the force. Fire fighter/EMT Denise Carver is 5 feet tall and weighs 120 pounds. Assistant Chief Woods has issued an old, small set of turnout gear to fire fighter Carver. This turnout gear does not meet the new department standards. He refuses to order fire fighter Carver a new set of turnout gear because, "Companies probably don't even make gear for women, especially ones that small." All other members of the department, including both of the other women, were issued new turnout gear whenever the department did not have gear available that met the new standards and fit the fire fighter to whom it was issued. When asked by Chief Marks about the fact that Carver was never issued comparable gear, Assistant Chief Woods says, "She doesn't need it anyhow. I don't want her going into a burning building: she might hurt herself."

1. Is this a form of sexual harassment?
2. Is this a form of sexual discrimination?
3. What recourse does fire fighter Carver have?
4. What should Chief Marks do?

Case #4

A small, full-time fire department of fifty-six officers and fire fighters has several female fire fighter/EMTs. The fire chief is experiencing marital problems. He alludes to this one day during a conversation with fire fighter Evelyn Davis. She expresses sympathy, saying, "That is really too bad. I know this is a rough time for you. I hope everything works out okay."

Two days later the chief approaches Davis, who is checking out equipment in the bay area, and tells her he had not noticed before how pretty and caring she is. Later that week, he touches her hair and tells her how much he is attracted to her. Davis begins to avoid the chief whenever possible and tries to make sure other fire fighters are around whenever he is there.

After a public education program, while Davis is packing up the last supplies, the chief tells the other personnel to return to the station and he will take the last few boxes in his car and give fire fighter Davis a ride back to the station. Finishing up, Davis discovers the EMTs have returned to the station and the chief is talking with the school principal about the program that was just presented. Not wanting to cause a scene, she accompanies the chief to his car.

On the way back to the station, the chief attempts to hold her hand and tells Davis he thinks he's falling in love with her. He asks her to go to dinner with him, "Just to talk, because talking to you makes me feel so much better." Davis declines and asks him not to make such advances to her anymore. When he reaches for her hand again, she jerks away and says, "Don't touch me! I don't want you to touch me! *Ever!*"

A few days later, the chief makes snide remarks to Davis and questions her ability as a fire fighter. After 2 weeks of treatment ranging from fierce verbal reprimands for the performance of her duties to giving small gifts such as flowers and candy, Davis realizes she hates to go to work because she dreads what may happen. No one else on the department has said anything about this conduct.

Davis decides to go to the safety director, who is supervisor to the fire chief, with her complaint and asks an attorney friend to accompany her. After telling the safety director about the situation, he responds, "Oh, you're just too cute." He then agrees to look into the matter, but says he thinks she is probably exaggerating the situation.

1. Who might be liable if a lawsuit is filed? Why?
2. Which of the chief's actions were inappropriate?
3. Did this start out as a case of sexual harassment?
4. If not, when did it become one?
5. Who was at fault? Why?
6. As the public safety director, what action would you take?
7. As the EMS coordinator, Davis' superior who reports to the chief, what action would you take?

Follow-up

The safety director spoke to the chief, who denied the allegations. The safety director took no further action. The chief began to issue unfair job assignments to fire fighter Davis and to issue formal reprimands for performance. Davis filed suit against the fire chief, the EMS coordinator (her immediate supervisor), the public safety director, and the city.

ENDNOTES

1. "Pay for Time Not Worked," in *Public Administration Review,* January/February 1989.
2. William Peterson, "Fire Department Administration and Management," in *Fire Protection Handbook,* 17th ed. (Quincy, MA: National Fire Protection Association, 1991), pp. 9–47.
3. For example, see *The Fire Protection Handbook* and *SARA Title III: Intent and Implementation of Hazardous Materials.*
4. Copies of the U.S. Code of Federal Regulations can be found in the government documents sections of most libraries. Copies may be purchased from the Government Printing Office, Washington, DC.
5. Many OSHA regulations have been built on established standards, including the fire codes developed by the National Fire Protections Association. For example, the OSHA *Fire Protection and Means of Egress* standard references NFPA *101, Life Safety Code,* for means of egress planning and safe exiting procedures. NFPA standards 10, 12, and 17 are referenced for fire detection, suppression, and extinguishing information, and planning guidance. Familiarity with NFPA standards is advised.
6. This discussion is based primarily upon the following paper: David H. Hoover, "U.S. Regulations on Hazmat Emergency Response" Chemical Disaster Workshop. Taiwan, R.O.C., November 1991. In Volume II of the Conference Proceedings.
7. 243 U.S. 210 (1917) Section 2.6 as cited in Warren Freedman, *The Law and Occupational Injury, Disease, and Death* (Westport, CT: Greenwood Press, 1990), p. 8.
8. Melissa A. Berman, "Talking through the Glass Walls," in *Across the Board,* July–August 1988, p. 26.
9. Steven W. Hays and T. Zane Reeves, *Personnel Management in the Public Sector* (Boston, MA: Allyn and Bacon, Inc., 1984), pp. 442–444.
10. Marilyn Loden and Judy B. Rosener, *Workforce America! Managing Employee Diversity as a Vital Resource* (Homewood, IL: Business One Irwin, 1991), p. 201.
11. Stanford University Self-Study on Building a Multiracial Multicultural University Community, Stanford University, April 1990, pp. 1–4.
12. Loden and Rosener, *Workforce America,* p. 73.
13. David J. Cherrington, *Personnel Management: The Management of Human Resources* (Dubuque, IA: Wm. C. Brown Publishers, 1987), pp. 485–487.
14. Jerry Holland, Vice President, 8th District, International Association of Fire

Fighters, AFL-CIO-CLC, Interview at the University of Akron, Akron, OH, August 1992.

15. Hays and Reeves, *Personnel Management,* p. 161.

REFERENCES

Cherrington, David J. 1987. *Personnel Management: The Management of Human Resources.* Dubuque, IA: Wm. C. Brown Publishers.

Cox, Taylor, Jr. 1991. "The Multicultural Organization." *Academy of Management Executive.* Vol. 5, No. 2. pp. 34–47.

Decker, Kurt H., and H. Thomas Felix, II. 1991. *Drafting and Revising Employment Handbooks.* New York: Wiley Law Publications.

Edwards, Audrey. 1991. "Cultural Diversity: The Enlightened Manager—How to Treat All Your Employees Fairly." *Working Woman.* Vol. 16, Iss. 1, January. pp. 45–51.

Gardenswartz, Lee, and Anita Rowe. 1989. "The Multi-Cultural Workforce." *Working World.* May 16, p. 16; June 26, p. 34; August 14, p. 20.

Hays, Steven W., and T. Zane Reeves. 1984. *Personnel Management in the Public Sector.* Boston, MA: Allyn and Bacon, Inc.

Holland, Jerry, Vice President, 8th District, International Association of Fire Fighters, AFL-CIO-CLC. Akron, OH: The University of Akron. Personal Interview. August 1992.

Hoover, David H. 1991. "U.S. Regulations on HazMat Emergency Response." *Proceedings of the Chemical Disaster Workshop,* Volume II. Taiwan, R.O.C. November 1991.

Johnston, William B., and Arnold E. Packer. 1987. *Workforce 2000: Work and Workers for the Twenty-first Century.* Indianapolis, IN: Hudson Institute, Produced for the U.S. Department of Labor.

Kern, Marilyn. 1989. "Minorities 2000." *Public Relations Journal.* Vol. 45, Iss. 8, August. pp. 14–18, 21–22.

Laporte, Suzanne B. 1991. "Cultural Diversity: The Sting of the Subtle Snub." *Working Woman.* Vol. 16, Iss. 1, January. pp. 53–55.

Loden, Marilyn, and Judy B. Rosener. 1991. *Workforce America: Managing Employee Diversity as a Vital Resource.* Homewood, IL: Business One Irwin.

Morrison, Ann M., and Mary Ann Von Glinow. 1990. "Women and Minorities in Management." *American Psychologist.* February. pp. 200–208.

Njeri, Itabari. 1989. "Intercultural Etiquette: When Different Groups Converge, the Ignorant and the Curious Can Be Unexpectedly Rude." Los Angeles Times. April 2, Part IV, p. 1.

"Pay for Time Not Worked." *Public Administration Review.* January/February 1989.

Sue, Derald Wing. 1991. "A Model for Cultural Diversity Training." *Journal of Counseling and Development.* Vol. 70, Iss. 1, September. pp. 99–105.

Zack, Arnold. 1991. *Grievance Arbitration: Issues on the Merits in Discipline Discharge and Contract Interpretation.* New York: Lexington Books.

CHAPTER 5

Hiring Practices

Hiring is a management function that has tremendous long-term impact on the operation of the fire department. Hiring and personnel choices are some of the most important decisions facing fire service administrators. Proper hiring practices can secure the needed talent and skills that the complexity of services offered by the fire service demand. There are many areas of expertise in today's fire service. It is important to know exactly what an employee is to do before hiring.

This chapter begins with a thorough discussion of the process of conducting a job analysis. A position is examined to determine what specific skills are necessary to meet the performance requirements of the job. After the job has been carefully analyzed, a meaningful job description focusing on the accomplishments desired is written.

The character and usefulness of a results-oriented job description make it one of the most useful tools for fire service officers in working with personnel. This chapter outlines how to design and structure a job description and provides a detailed example.

Techniques for recruiting new members for the fire service will also be discussed. Topics include variations among types of departments and recruiting minorities. The special approach to recruiting and hiring executive fire officers receives separate discussion.

The role of testing and methods of validating qualifications tests are issues receiving increasing attention in the fire service. Questions must be relevant to the job and validated through an accepted methodology. In addition, testing must be unbiased and uniform.

This chapter will assist fire officers addressing different aspects of the recruitment and hiring process. Fire officers who are not in the process of hiring can use the information about job analysis and results-oriented job descriptions to modify and streamline existing staffing assignments. These techniques can increase both the efficiency and the effectiveness of the fire service.

JOB ANALYSIS

Before someone is hired, the organization/fire department must identify exactly what the individual is to do. Job analysis is the process of identifying what activities are to be performed, what skills are required, what knowledge is necessary, etc. The information gathered in a job analysis is used to write a job description, which is a statement detailing the work performance expectations of the position.

Knowing exactly what is expected from an employee is important in various areas of administration. As we look at the aspects of performing a job analysis, we find that the information collected can be used in a host of ways, for example:[1]

- Personnel planning
- Recruiting
- Selection
- Orientation
- Evaluation
- Compensation
- Training
- Discipline
- Safety
- Job redesign

The investigator gathers all available information about a job and identifies what knowledge, skills, and abilities (KSAs) are required to perform the job functions. The investigator must first define exactly what tasks are performed in a specific job, then identify what one needs to know to perform these tasks. For example, a fire inspector must, among other things, be familiar with the local building and fire codes. This is a specific area of knowledge. A fire inspector must make on-site inspections, a task that requires the ability to access various parts of a building and examine pertinent areas. Although this seems simplistic, identifying exactly what is required in a given position is the first step in job analysis. A number of techniques can be used to conduct this job analysis.

Observation involves visiting the place of work and watching an

FIGURE 5-1 Observations and interviews can help determine the specific tasks associated with a given position.

individual perform his or her tasks. This is difficult in the fire service given the variety of tasks performed, the uncertainty of each day's schedule, and the risk involved in emergency response. However, it can be included as one component of a job analysis. Observation permits a double-check of reported percentages of time spent on various components of a job. For example, if a company claims that its members spend a minimum of 20 hours a month in training and, in fact, only 10 hours of training are observed for two different months, the claim can be challenged and the job description or assignments altered.

Interviews with current fire fighters and supervisors can be useful because they can be scheduled away from the actual work location and do not interfere with daily operations. If designing a new position, it is necessary to interview the supervisors of the new position and advisable to interview others who will be interacting with a new fire fighter. In addition, if the new position is comparable to one in a different community, it is a good idea to interview the current person holding that job as well as his or her supervisor.

For example, if a fire department is going to add the position of public education specialist, the investigator should interview the fire prevention officers, the fire chief, fire fighters who currently conduct public education programs, and full-time public education specialists and their supervisors who work with fire departments in other communities. These interviews focus on what the job entails and what the individual needs to know and be capable of doing to perform the required task. The main focus of the interview should be task identification.

Expert panel task analysis is another way to identify the tasks performed in a given position. A group of experts in the field are gathered and asked to identify what a person in the position in question does. Once this stage is complete, they are then asked to identify, for each task, what areas of knowledge, skills, and abilities are needed to accomplish that task. These form the basis for the job description.

This is a detailed, time-consuming process. Once it is done, however, identifying the job candidate's qualifications is much easier. In fact, when developing NFPA 1021, *Standard for Fire Officer Professional Qualifications,* a task analysis was performed on the various officer levels and used to derive the standard. Job analysis is a useful technique when setting standards, whether they are national standards on officer qualifications or local department standards for a given officer or fire fighter position. (See page 123.)

The committee developing NFPA 1021 first identified what officers at each level were expected to do and what duties they might be called upon to perform. Then the committee identified the areas of knowledge necessary to perform those duties and the administrative skills required. This information was used to define the various levels in the standard. For example, for each officer level the degree of familiarity with budgeting and financial resources changes. This is because the requirements range from making only allowable purchases to preparing a budget to assisting in determining what financial resources the community is able to provide. Using job analysis to develop performance expectations for each officer level provided a basis for consensus among committee members and resulted in matching performance requirements and characteristics for each officer level. This consensus about required qualifications and performance criteria is crucial to the

EXPERT PANEL TASK ANALYSIS FOR PUBLIC EDUCATION SPECIALIST

What does a public education specialist do?

Lectures/gives talks
Designs programs
Works with children
Writes reports
Makes posters
Secures handout materials from fire organizations and agencies
Etc.

TASK: LECTURE/GIVE TALKS
What must the public education specialist need to know to lecture?

Fire safety

What skills must the public education specialist have to lecture?

Public speaking

What ability must the public education specialist have to lecture?

Speak clearly
Communicate well
Communicate with different age groups

TASK: DESIGNS PROGRAMS
What knowledge must the public education specialist have to design programs?

Education methodology
Course design
Fire safety

What skills must the public education specialist have to design programs?

Design lesson plans
Write presentation material
Prepare visual aids

What abilities must the public education specialist have to design programs?

Organize information
Conceptualize visual aids
Adjust level of information to audience
Etc.

success of results-oriented job descriptions. These carefully established criteria define the scope of a job, a means to evaluate job performance, and the qualifications required for the position.

Task analysis is not something that should be done quickly or haphazardly by an inexperienced person. It is a serious activity that should be conducted by someone with background and training in the field. In addition, all of the participants must be totally objective about the position in question. Tailoring the analysis to a particular individual's qualifications does not provide the fire service, the department, or the individual an accurate picture of the scope of the position.

Questionnaires can also be used to gather information. They are generally in two parts. The first section asks the worker to describe the kinds of experiences, qualifications, attitudes, information, and skills necessary to perform the job. The second section contains a list of various activities. The worker is asked to identify those activities performed on the job and what percentage of time is spent on each. The Position Analysis Questionnaire (PAQ) is one of the best-known analysis questionnaires.[2] It consists of 194 different job elements that measure six categories of a job: (1) information input, (2) mental processes used, (3) work outputs, (4) relationships with other workers, (5) job context and work satisfaction, and (6) other job characteristics. The information from this questionnaire is then used to classify the various job dimensions. In addition, the information can be matched with information about jobs to identify similarities in worker abilities required. This type of questionnaire is also available for managerial positions. [Two examples are the Managerial Position Description Questionnaire (MPDQ) and the Executive Position Description Questionnaire (EPDQ).]

It is not certain how applicable these standard questionnaires are for fire departments. The fire service requires considerable career-specific technical knowledge that is not required of administrators in either the general business or public sectors. For example, a fire officer primarily concerned with budgeting must be aware of department standard operating procedures that dictate the type and size of hose used most frequently. Someone charged with preparing bid specifications must be familiar not only with the city's purchasing process and restrictions associated with expenditure approval, but also with the safety standards and technical requirements of the equipment in question. Many business and industry professionals believe, as many members of the fire service do, that they have a unique position or set of positions not adequately described or defined by the general management questionnaires, and they hire personnel analysts to design questionnaires specifically for their organization. These are used to define and refine the job description and classification. This is one option fire departments could consider. In fact, if a regional fire service organization were to work on such a project together, each fire department could benefit from the work of a consultant who could design general position questionnaires for the fire service.

Employee recordings, i.e., having employees maintain a written record of what they do, are another means of collecting job information. These are usually done in 15- or 30-minute increments. When using this technique, it is vital to record the information on a continuing basis. Recording the information at the end of the day does not result in the kind of information necessary to perform a job analysis. The information gathered from such recordings can be used to identify specific tasks performed and associated skills required. This technique is often used to double-check questionnaires and interviews. As many administrators know, responsibilities that comprise 10 percent of the job description often take 50 percent of the time—or more. Such information can be documented through the use of employee recordings.

The data received can be used to change a job description or to argue for the need of a new position. For example, in some fire departments the responsibility for recording and processing SARA Title III information on the presence of hazardous materials was initially assigned to the fire inspection division because the first step in pre-incident fire planning was a function of the prevention bureau. The time required to comply with SARA Title III, however, became so overwhelming in communities with a larger industrial base that the inspection bureau could not perform both functions with the same level of staffing. Employee recordings documented the number of hours spent processing the hazardous materials information and provided the justification for new positions.

Once the information on the job analysis is collected, it is grouped into logical categories based on required performance of tasks and associated qualifications necessary to perform these tasks. Job descriptions are constructed using this information.

JOB DESCRIPTIONS

Job descriptions are written statements detailing an employee's duties precisely. Most traditional job descriptions also include a qualifications standard identifying the minimum qualifications necessary to perform the duties specified and a statement indicating to whom the position reports. Traditional job descriptions are often of limited value, however. They cannot be used to focus training, evaluate relative merits of job candidates, or tell the job candidates about work conditions and expectations.

The need for more useful job descriptions led to several improvements. "Results-oriented job descriptions" were so named because they focus on the outcome or expected performance standard of the employee. This is similar to the training and education objectives utilized by many fire service programs, including the National Fire Academy. The National Fire Academy develops learning objectives for each unit in its courses. These are based on the ABC objective, which is taught in their Educational Methodology course: A stands for the actor, B for the behavior, C for the conditions. Thus, a training objective for teaching rookie fire fighters proper use of SCBA might state:

"The fire fighter	(actor)
will correctly put on an SCBA	(behavior)
in a smoke-filled room in 30 seconds."	(condition)

Klingner and Nalbandian expanded on this format in discussing the results-oriented job description.[3] The job descriptions they developed have five sections each: Tasks, Conditions, Standards, KSAs, and Qualifications.

These provide a much clearer explanation of what is expected of the individual in the job being described. Results-oriented job descriptions can also be used for training and evaluation, and can be referenced in disciplinary procedures. Information gathered from evaluating performance standards may indicate a problem in the basic structure of the job function. The job function can then be restructured and the performance standard modified.

The following figure demonstrates what a results-oriented job description for a fire inspector might look like. As you review this results-oriented job description, identify what aspects you would change. Then identify what job analysis techniques you could use to support the changes.

JOB TITLE: FIRE INSPECTOR

Tasks	*Conditions*	*Standards*
• Inspect industrial buildings	• As assigned by fire prevention captain by area of responsibility	• Each assigned structure inspeted each year/ all violations corrected annually. Reports used In preplanning.
• Inspect commercial buildings	• As assigned by fire prevention captain by area of responsibility.	• Each assigned structure inspected every 18 months. Violations corrected annually. Reports given to planning officer.
• Inspect places of public assembly	• As necessary to assist other inspectors at direction of fire prevention captain	• Each assigned structure inspected within 30 days of assignment. Violations corrected 30 days thereafter. Report shared with other inspectors.
• Approve new building plans	• Preview and approve building plans assigned by the fire prevention captain	• Work with building inspector to review plans within 2 weeks. Negotiate changes within 3 weeks. No complaints from developers.

Tasks	*Conditions*	*Standards*
• Approve new construction of industrial, commercial, retail, and places of assembly	• Minimum of three on-site inspections during construction plus one final inspection upon completion	• Visit site during appropriate stages of construction to view structural supports, fire stops, electrical equipment, and installation of fire detection and suppression systems. Provide plan of site to planning officer. Good review from builders.

Skills, Knowledge, and Abilities Required

Ability to conduct on-site inspections
Ability to read and interpret blueprints
Knowledge of fire and building codes
Knowledge of fire protection concepts and equipment
Ability to work with members of other departments
Ability to work with developers
Ability to work with builders
Ability to communicate
Ability to write reports
Knowledge of preplanning information

Minimum Qualifications

High school diploma or equivalent
Four years experience in fire inspection
Hold current state certification as a fire inspector
Possess a valid driver's license

This job description provides a clear understanding of exactly what is expected of the fire inspector and what the working conditions will be. In addition, tying the KSAs and the minimum qualifications directly to the tasks to be performed clearly demonstrates why those qualifications are necessary. This eliminates the possibility of including capricious qualifications that may discriminate against certain applicants, such as minority groups, members of certain religions, women, or even applicants from outside the community.

A results-oriented job description can help employers identify the most qualified applicants for the position. Interviews can focus directly on the various aspects of the job description. Reasons for rejecting candidates should relate directly to that description as well.

Once employed, if an employee fails to meet performance standards set in the job description, an officer has grounds for disciplinary action. Because the job description is clear, both the fire officer and the fire fighter know exactly what is expected, what is acceptable, and what is not. The standards in the job description can also be used as goals for employees who are not working at an acceptable performance level. These can be the focus of employee development activities tied to disciplinary actions. (See Chapter 6, Motivation and Discipline, for further discussion.)

The one drawback of a results-oriented job description is that as conditions change, the job description must also change. For example, a sudden spurt of construction activity may not allow the current inspectors to visit each commercial structure every 18 months, and blueprints may not be reviewed in 2 weeks' time. When this occurs, the standards must be adjusted. Likewise, if a job is expanded in some way, the job description must also be revised.

Job expansion refers to instances where the duties and tasks of a specific job are increased by expecting more of the same type of activity (inspecting 75 businesses instead of 50) or by adding responsibility for new activities (preplanning industrial facilities upon completion of inspections). Job expansion can be part of a comprehensive employee development and job enhancement process, or it can be a way of overloading someone unfairly. The first instance is a motivation technique that yields a positive outcome for the department. The second is an unfair practice that tends to overburden workers, lower morale, and decrease performance.

Job expansion and enhancement should be accompanied by a job analysis and a full and open revision of a job description. If there is someone currently in the position that is to be expanded, the proposed changes must be discussed with this individual. This is especially true if the description of the position or the duties are part of a labor agreement. Even if the proposed changes are not covered by contract, it is always advisable to discuss any such changes in task assignments with the fire fighter before making changes. If fire fighters have pride in their work and care about the people they serve, they will want to have a say in how their work is going to be changed. In addition, fire officers need to remember that no one knows more about a specific job than the person doing it. Officers can learn a lot by including fire fighters in discussions about work assignments and task allocations.

LABOR AGREEMENTS

Chapter 4, Human Resources Management, discusses the role of labor agreements in the administration of the fire service. These agreements must also be considered when modifying existing job descriptions, especially if the performance requirements are specifically mentioned in the agreement.

Any job analysis and change in job description covered by union labor agreements must take place with the consent, and usually the participation, of the union. Job performance, conditions of employment, and expectations are usually major components of labor agreements. Fire department administrators cannot arbitrarily change job descriptions or performance standards.

It is possible to work with the union to update job requirements and make them more realistic. In addition, job analyses and new job descriptions may be negotiated during the bargaining sessions. The fact that a department is under union contract does not mean job analysis cannot or should not take place. A job analysis can demonstrate needed changes. This information can be used to negotiate with the union for changes to improve working conditions and the overall operation of the fire department. It is still the primary duty of fire officers and administrators to provide the best service possible to the public. This means that even though tough negotiations may be involved, if changes need to be made, fire administrators must work with the union to see that they happen.

PUBLIC NOTICES

One of the primary uses of a job description is for hiring. In order to hire, the department must solicit applications. Therefore, the obvious first step to take is to advertise available positions.

Many fire departments have policies stating that current members of the department have the first opportunity to apply for any open positions. This encourages promotion from within and allows for career development in the department. In many combination departments this allows individuals who have spent years volunteering time on a part-paid basis to have the first opportunity to secure full-time employment as those positions arise. In these instances, the first posting of position notices should be made in-house.

When this posting is made, it is imperative that each member of the department be informed and have access to equal information. Providing more information to one individual or group of individuals than another can be interpreted as a violation of Equal Employment Opportunity (EEO) standards. Such sharing of information can be unintentional; even so, it is not excusable.

For example, a combination fire department was going to add a 40-hour-per-week fire inspector position. This was announced at the Tuesday night and the Wednesday morning training sessions. The chief stopped by during the Wednesday morning session and the fire fighters began to ask questions about the position. The chief talked with them for about a half hour and explained in detail what the department wanted and how the applications were going to be evaluated. No one who attended the Tuesday night meeting had direct benefit of this information and most never even heard about it secondhand. This gave one set of employees an unfair advantage over another set.

Specific requirements of EEO standards are discussed in Chapter 4. This discussion centers on the impact these standards have on hiring.

Postings outside of the department should be made in accordance with the laws of the government entity having jurisdiction over the fire department. This includes county, city, village, district, township, and state governments. Most states have laws stipulating the number of times personnel positions must be announced in the public sector. These laws apply to all public positions within the state. Many local communities also have similar regulations about advertising public positions. A fire department must abide by these policies.

Generally, the positions are advertised in commercial newspapers, trade magazines, and professional association newsletters. They are also posted in personnel departments and the public employment bureau. These postings should include the job description or a variation of it and clear instructions on how to apply for the position.

HIRING OR RECRUITING FOR VOLUNTEER DEPARTMENTS

It is true that individuals tend to join volunteer fire departments out of a sense of service to the community or civic pride, or because of a personal love of the fire service. The changing nature of today's society, with increased obligations and changing family structure, has reduced levels of volunteerism in all areas. In spite of the call for volunteerism throughout the 1980s, there was an actual decrease in the number of individuals volunteering their time to assist different community, social, civic, or religious organizations. Thus, it is not surprising that many volunteer fire departments are also experiencing difficulties recruiting members.

Legal mandates have increased required training time and set stringent safety standards for fire fighters, making it more difficult to recruit volunteers. It is one thing to ask volunteers to assist in emergency response and aid citizens of the community. It is another to require volunteers to spend increasing amounts of time and money on training before they are able to help someone in an emergency. This often deters prospective volunteers.

One effective recruiting technique for volunteer fire departments is to contact various civic organizations to inform them of the need for new members. For example, a presentation at a local community service organization may spur interest from attendees. In addition, these individuals will assist in spreading the word throughout the community.

The local newspaper and radio station can also assist the recruiting endeavor. A lead story about the staffing needs of the local fire department will reach most community residents. The news report can also provide information about the performance expectations so potential candidates have an idea of the job requirements.

Part-Paid Departments

Many fire departments are staffed by part-paid personnel or paid on-call personnel. These departments may also benefit from some of the techniques discussed for recruiting volunteer fire fighters. Recently, however, a number of paid on-call or combination fire departments have discovered that the newspaper stories that worked in the past have less impact than a simple advertisement for part-time help wanted. Many of the new recruits in part-paid departments are career fire fighters from other departments who want part-time work in another department. Other applicants have full-time jobs in another career field but have always wanted to be a fire fighter and recognize this as an opportunity to be one. Many of the new recruits in part-paid and combination departments view the position on the department as a part-time job rather than a voluntary commitment of time.

Regardless of the type of fire department, a variety of outreach and recruitment techniques will need to be used when seeking new members. Being familiar with the community can help fire service officers target specific areas of the community for recruitment and select the most appropriate and effective techniques. After all, a fire department must have applicants before it can hire. While this may not be a problem for larger departments, it is an increasing challenge for volunteer and part-paid fire departments.

MINORITY OUTREACH

The changing demographic composition of our nation was detailed in Chapter 3, Cultural Differences in the Community. The impact this change is having on the internal composition of fire departments was the focus of considerable discussion in Chapter 4, Human Resources Management. Minority outreach activity is an important and integral component of the hiring process that cannot be overlooked. The concepts of EEO and Affirmative Action, of increasing the representation of all ethnic groups of the community on the fire department, cannot be relegated to simple discussion. It needs to be a central consideration as job announcements are made and recruitment takes place. This section focuses on applying EEO and AA policies to the hiring process.

Much work has been devoted to identifying ways to encourage women and minorities to apply for positions in occupations where they are underrepresented.[4] One effective method of outreach is publishing the job announcements in local papers and magazines that address the interests of the target groups, including the statement "women and minorities are encouraged to apply." A number of metropolitan areas have weekly newspapers that are circulated to various minority groups.

FIGURE 5-2 Sample job advertisement with emphasis on recruiting minority candidates.

Fire Fighters Wanted

The City of Lakeville is seeking qualified men and women to apply and test for upcoming fire fighter classes. Will perform critical work in fire suppression and emergency medical service. Applicants must be 18 to 34 years old, hold valid Michigan driver's license and Lakeville residency, be in excellent physical condition, and possess excellent reading skills (foreign-language skills a plus). Paid training provided. Pick up applications at any Lakeville fire station; completed applications must be received by June 15. The City of Lakeville is an Equal Opportunity/Affirmative Action employer—women and minorities encouraged to apply.

In addition, there are publications that specifically target women's interests. Placing job announcements in these media can result in a number of minority and women applicants.

Another good outreach technique is to post the job notices in minority neighborhoods and public places, such as laundromats, grocery stores, and convenience stores. Any employment agency or commission that receives the job notice should be asked to encourage minority and female applicants.

PROCESSING APPLICATIONS*

A central and secure point for receiving the applications should be established as soon as the advertising begins. These applications are confidential and should not be made available to any person not directly involved in the hiring decisions. Each application should be reviewed to determine whether the applicant meets the minimum qualifications stated in the job description. Those who do not meet the stated qualifications are disqualified and removed from the selection process. Those meeting the minimum qualifications then go to the next step in the process. For many communities the next step is testing; for others, no testing is required. In departments without testing requirements, a careful system for reviewing and rating applications needs to be established before the applications are actually reviewed to ensure fairness and an objective measure for selecting candidates.

A rating system should be established for each qualification and for each area of knowledge, skills, and abilities identified in the job description. This system should have at least three categories: "Meets minimum qualifications," "Exceeds minimum qualifications," and "Superior qualifications." The different qualifications and, especially, KSAs should then be ranked in order of importance. Ranking applicants using a system of objective comparison provides a method of selecting the best candidates for the position.

*The following text and accompanying forms are excerpted from Corporate Research Inc., P.O. Box 1455, Akron, OH 44309. Do not use without permission of Corporate Research Inc.

EXPERIENCE QUESTIONNAIRE FOR POSITION OF ASSISTANT CHIEF STANDARD FIRE DEPARTMENT

This questionnaire has been designed to gather information to be used to evaluate your qualifications for the above position. This gives you an opportunity to provide information on your background and, in particular, your accomplishments. We are especially interested to know about things you have done. Please read the following instructions carefully.

1. A number of areas important to the Position of Assistant Fire Chief in the Standard Fire Department have been identified. For each one, in the spaces provided, fully explain your experience in this area and what you consider to be your two greatest accomplishments. Identify, also in the locations provided, the person who can best provide additional information on your accomplishments.

2. Your answers should be specific and concise so that the readers know exactly what you have done. You may find it helpful to consider the following questions as you write your answers:
   a. What was the situation or challenge you faced?
   b. How did you approach the problem? What did you do?
   c. What was the result, or impact, of your activities?

3. It is possible you may not have sufficient space to answer in the detail you think appropriate. You may attach additional pages, which must be stapled to this booklet. Be certain to indicate the item number which is being continued.

4. Your answers must be printed in black ink, or be typewritten. Legibility is important.

5. Your answers are subject to further inquiry and verification.

6. Think carefully about your responses. You may wish to prepare an outline before you begin writing.

7. An example appears on the following page.

EXAMPLE

Dimension 23:
FIRE FACILITY MAINTENANCE

During my five years as Captain, my responsibilities included serving as shift captain and supervising the fire inspection program. In 1988, I directed a survey to identify fire station maintenance requirements. We wanted to identify:

1. immediate maintenance needs, and
2. anticipated requirements for the next 5 years.

The survey was a written document about 10 pages in length, and it was distributed to all members in the ranks of Lieutenant and higher. (The response rate for this survey was 94%.)

I analyzed the results and briefed the Fire Chief. The Assistant Chief and I then developed a plan to meet the required maintenance.

We tracked the costs for facility maintenance over the last 6 years and found out that the cost of maintenance has decreased and the level of cleanliness and readiness has increased.

In addition, during labor contract negotiations the year before this program was begun, the union's representative complained repeatedly about "living conditions" in the stations. The Chief received periodic complaints of the same nature. No such complaints were received after the first year of the maintenance program. Employee morale improved.

My second accomplishment was a cost savings in the procurement of supplies. I learned that the Fire Division was not taking advantage of the lowest prices available on expendables, such as cleaning supplies, paint, and repair hardware. I asked a Lieutenant to work with the Village's Finance Director to identify the lowest cost source for supplies. As a result, the cost for supplies was reduced 15% over the first year.

With which employer was this experience obtained?
Smithville FD

When? 02/79-01/88

Who is most knowledgeable about your accomplishments?
Chief Thomas E. Magnuson, Jr.

Full title: Chief, Smithville Fire Division

Phone: 513/555-1212

Dimension 1:
DEVELOPMENT OF STANDARD OPERATING PROCEDURES

With what employer was this experienced obtained?

____

When? ________________________

Who is most knowledgeable about your accomplishments?

____

Full title: ________________________

Phone: ________________________

Dimension 2:
PUBLIC FIRE AND SAFETY EDUCATION PROGRAMS

With what employer was this experienced obtained?

____

When? ________________________

Who is most knowledgeable about your accomplishments?

____

Full title: ________________________

Phone: ________________________

Dimension 3:
FIRE DEPARTMENT PLANNING

A number of different dimensions can be used. In this example, there were a total of eight. The remaining ones were:

4. Direction and/or performance of fire cause and origin investigation
5 Supervision and conducting of training activities for fire, EMS rescue, & administration
6. Budget preparation and oversight including maintaining accounts of bills and purchase orders
7. Equipment and apparatus maintenance and purchase
8. Employee supervision

A separate sheet was provided for each of these dimensions as well.

With what employer was this experienced obtained?

____

When? ________________________

Who is most knowledgeable about your accomplishments?

____

Full title: ________________________

Phone: ________________________

Each of the dimensions was rated according to its degree of importance (8 = most important, 1 = least important) to the Fire Chief and members of the governing board (this could be the Township Board of Trustees, City Council, or Safety Director) or hiring and promotion board. These ratings were used to weight the relative importance of the different activities or dimensions. A score sheet was then developed. Each candidate was rated on a scale of 1–10 with 1 representing the least amount of demonstrated accomplishments and 10 being the greatest amount of demonstrated accomplishments.

The scores are then entered onto the rating sheet. A copy of this form follows. The weight for each dimension is already entered. One rating sheet is completed for each candidate by each evaluator. We recommend that at least three individuals evaluate the applications independently.

The score for each is then computed by multiplying the score for each dimension by the pre-assigned weight. These values are then added together and the sum is the total accomplishment score for the candidate given by the evaluator. The evaluators then compare their evaluations for the highest scoring candidates. If everyone agrees on which dimensions are most important, the top five candidates should be the same for each evaluator. These applications can then be carefully reviewed and interviews can be scheduled for the top scoring three to five candidates.

Candidate: ______________________________

ASSISTANT CHIEF—Standard Fire Department

Dimension	Score	x Weight	= Weighted Score
1. Development of SOPs	_____	4	_____
2. Public Fire and Safety Education Programs	_____	3	_____
3. Fire Department Planning	_____	7	_____
4. Direction and/or Performance of Fire Cause and Origin Investigation	_____	2	_____
5. Supervise & Conduct Training Activities for Fire, EMS Rescue, and Administration	_____	5	_____
6. Budget Preparation and Oversight Including Maintaining Accounts of Bills and Purchase Orders	_____	6	_____
7. Equipment and Apparatus Maintenance and Purchase	_____	1	_____
8. Employee Supervision	_____	8	_____
TOTAL SCORE			_____

Perform any verification of credentials and background checks that are necessary at this point. Schedule an interview with the top candidates. Any major problems identified during the interviews should be taken into consideration during the final employee selection. It is advisable to have more than one person interview the final candidates in order to draw a more objective opinion.

Because of the EEO requirements and local and state mandates, it is necessary to document each step in the hiring process. The applications of all candidates must be kept together with documentation of their disposition. In addition, to meet EEO and AA requirements the department should keep a record of the number of minority and female applicants there were; why rejected minority applicants were disqualified; how far they progressed in the process; whether there were any women or minority finalists; and, if so, why those rejected were not hired.

TESTING

After the initial screening, many fire departments rely on tests to determine eligible hiring pools and to rank the applicants. If testing is part of the hiring or promotion process, it must be conducted for *each* position open *every time*. Testing policies that enable the fire chief or other officers to test on some occasions and not on others are capricious, discriminatory, and illegal.

There is more to giving a test than simply writing up a few questions or picking questions out of a book. Tests of this nature have been successfully challenged in court as being unrelated to the performance requirements of the job. The test must deal only with the KSAs identified in the job description. Anything beyond that is considered irrelevant to the performance of the job and an unnecessary requirement.

Test Validation

Test validation is the process that determines to what extent a test is related to the job. Moreover, test validation reveals the degree to which the test measures those aspects of the job it is supposed to measure. An invalid test for hiring or promotion can be successfully challenged in court as a biased and capricious activity irrelevant to the performance of the job and used to discriminate against individuals or groups of individuals. The Uniform Guidelines on Employee Selection Procedures of the EEO[5] require that any test that may affect affirmative action groups must be validated. In addition, members of non-affirmative action groups may also challenge tests unrelated to job performance. Thus, fire departments must conduct test validation when conducting new tests for promotion or hiring. This may require using a qualified test specialist if the fire department does not have personnel trained in this field.

There are three primary methods of test validation: (1) criterion-related validity, also known as empirical validity, (2) construct validity, and (3) content validity.

Criterion validity is based on the mathematical correlation between test scores and subsequent job performance using statistical evidence. One way to measure this is to compare test scores with performance evaluations. For example, a test might be divided into three or four parts, relevant to various aspects of job task requirements. Performance evaluation scores in these areas would then be compared with the test scores. If a significant correlation is found for one part of the test and not the others, the first would have achieved criterion validity but the others would not and could be challenged as a valid and reliable criterion for employment.

Looking at the job description for a fire inspector, we can examine two required KSA areas in order to illustrate how this might occur: test section A (20 questions) to determine ability to read blueprints and test section B (20 questions) to determine ability to work with developers.

Applicant Name	*Test Score A (out of 100)*	*Test Score B (out of 100)*	*Performance Score A (out of 10)*	*Performance Score B (out of 10)*
Allen	70	80	7	6
Jones	45	64	4	6
Smith	90	85	7	9
Denniss	95	45	10	7
Martinez	85	50	9	10
Wyatt	60	85	6	4
Kensey	55	40	6	7
Tyler	80	80	8	4

Section A test scores correlate with performance scores, indicating that the test score predicts how well the person will read blueprints on the job. However, for section B, the scores measuring ability to work with developers, there is no statistically significant correlation. This portion of the test does not predict performance and is, therefore, invalid.

Applying statistical techniques, such as multiple regression analysis, will establish empirical or criterion validity and enable test specialists to set minimum acceptable test scores that guarantee, at least to the point of high statistical probability, that fire fighters who score above that level on the test will perform satisfactorily in the job.

Construct validity is the method used for most psychological testing. It involves identifying traits or aptitudes that are related to either good or bad job performance and then measuring these traits in job applicants. Applicants who demonstrate a high level of desirable traits or aptitudes are hired and those who demonstrate a high level of negative traits or aptitudes are immediately disqualified. Those in the middle are generally placed in a second choice category.

Tests involving psychological predictors and construct validity require the participation of qualified occupational psychologists. Positive traits are often identified by testing successful individuals currently in the positions in question. This requires a psychologist's education, familiarity with tests, and experience. As a result, this type of validity is primarily reserved for aptitude and trait testing of personality, which are part of the psychological profiles required by some of the larger fire departments.

Content validity is based on the ability to prove that the KSAs being tested are, in fact, part of the job. For example, given the job description, it is appropriate to test a fire inspector's ability to read blueprints; it is not appropriate, however, to test knowledge of fire suppression or fire modeling techniques. Neither of these two knowledge areas are identified in the KSAs or job qualifications. Had the job description included the qualification "be a trained fire fighter," or the task description "answer fire suppression calls for all second alarm fires," a test of fire suppression knowledge and abilities *would* be appropriate.

A detailed job analysis is necessary when using content validity.

The fire department or other test giver must be able to demonstrate that each KSA tested is necessary for the satisfactory performance of duty. This form of validity can be developed by fire departments, but it is a painstaking process that requires careful attention to detail and total objectivity.

Test Subject Restrictions

For both entry-level and promotional positions, it is only possible to test for knowledge the individual is expected to have *prior to any training that is part of the job.* Thus, if fire fighters will be given paramedic training once they are employed, they cannot be tested on any medical treatment techniques. Likewise, if the department has an officer orientation training class for all new officers, nothing that is taught in that class can be on the exam. The function of a test is to evaluate knowledge and abilities that the individual must have prior to taking the job or that indicate the individual's ability to learn the necessary skills in the training component.

Consistent Testing Policy

Testing cannot be a capricious requirement that fire officers or public administrators enforce inconsistently. In fire departments not covered by civil service, the practice of testing is left to the discretion of the fire officers and other local government administrators. Testing is not mandatory, but it is important that if a decision to test is made, it be done on a regular basis.

Civil Service Departments

Fire departments covered by civil service requirements must test for positions. Most of these tests are conducted by the state or by state-approved testing services. These tests have been validated, and the job classifications are based on "usual and customary" classifications and descriptions used in a wide variety of fire departments. Some local governments have their own civil service system and do in-house testing with either their personnel department or with approved testing

FIGURE 5-3 Testing for entry-level and promotional positions must be confined to knowledge the candidates are expected to possess prior to any training those personnel will receive on the job.

companies or individuals. In these cases, it is simply a matter of costing out and contracting for the services.

The testing services may or may not include advertising the positions as required by state, federal, and local laws. Even if the fire department contracts out the advertising, it is still responsible for compliance with the law. Thus, it is important to either contract with a company with an established track record, monitor advertising activities, or secure some type of bond or guarantee of compliance.

Screening for Executive Officer Positions

Qualifications for executive fire officers are more complex than those of rookie fire fighters. Abilities necessary to perform the job include the ability to motivate, to communicate complex technical information to public administrators, to negotiate resources, to budget, to plan, etc.

As a result, these promotions and hirings benefit from an approach that goes beyond simple testing. Qualifications such as the ability to differentiate between types of orders or verbal commands can be measured with a written test. Other qualifications, such as describing how understanding people and motivating department members affects group behavior in the organization, cannot be easily measured by standard written tests.

INTERVIEWS

Interviews are one common component of hiring for senior officer positions. These interviews should follow a set format and should be the same length for each candidate. Asking one set of questions of one

SAMPLE INTERVIEW GUIDE FOR CAPTAIN IN CHARGE OF EQUIPMENT MAINTENANCE AND REPLACEMENT

Describe your experience in overseeing equipment maintenance.

- Did you have an ongoing preventative maintenance program in place? (If so, please describe it.)

  (If not, why not?)

- How would you implement a preventative maintenance program in this department?

Where does the accountability for reporting equipment problems and responding to these reports lie? How do you monitor accountability?

Which NFPA standards are most relevant to equipment maintenance? [This question is asked to determine familiarity with various professional standards relevant to equipment maintenance. It demonstrates understanding of the field.]

Have you ever written bid specifications? For what equipment?

How did you determine the relevant specifications for the equipment?

How do you involve fire fighters and other officers in the equipment selection process? [This question offers an insight into leadership style.]

Have you worked with a periodic replacement and upgrade program for loose or hand-held equipment? Describe the program.

candidate and a different set of another candidate may not only give one candidate an advantage, but it also makes it impossible to compare the candidates on the same areas of qualification. For example, if an interview with Assistant Chief Peters deals with fire suppression techniques and fixed systems and an interview with Captain Myers deals with budgeting and resource allocation, there is no way to compare the relative capabilities of the two candidates for the same position.

The questions to be asked in the interviews should relate directly to the job assignments for which the officer will be responsible. See Table 5-1 for examples of questions relevant for a captain in charge of equipment. Asking about qualifications or experience that has nothing to do with the duties of the job is as inappropriate as putting such items on a test. The chart on the following pages provides a guide for the types of questions that are appropriate under EEO/AA guidelines.

TABLE 5-1 The Employment Interview Guidelines

Inquiries Before Hiring	*Lawful*	*Unlawful**
1. Name		Inquiry into any title which indicates race, color, religion, sex, national origin, handicap, age or ancestry
2. Address	Inquiry into place and length of current address	Inquiry into foreign addresses which would indicate national origin
3. Age	Any inquiry limited to establishing that applicant meets minimum age requirement that may be established by law	A. Requiring birth certificate or baptismal record before hiring B. Any inquiry which may reveal date of high school graduation C. Any other inquiry which may reveal whether applicant is at least 40 and less than 70 years of age

TABLE 5-1 The Employment Interview Guidelines (continued)

Inquiries Before Hiring	*Lawful*	*Unlawful**
4. Birthplace or national origin		A. Any inquiry into place of birth B. Any inquiry into place of birth of parents, grandparents or spouse C. Any other inquiry into national origin
5. Race or color		Any inquiry which wouldindicate race or color
6. Sex		A. Any inquiry which indicate sex B. Any inquiry made of members of one sex, but not the other
7. Height and weight	Inquiries as to ability to perform actual job requirements	Being a certain height or weight will not be considered to be a job requirement unless the employer can show that no employee with the ineligible height or weight could do the work
8. Religion—creed		A. Any inquiry which would indicate or identify religious denomination or customs B. Applicant may not be told any religious identity or preference of the employer C. Request pastor's recommendation or reference

TABLE 5-1 The Employment Interview Guidelines (continued)

Inquiries Before Hiring	*Lawful*	*Unlawful**
9. Handicap	Inquiries necessary to determine applicant's ability to substantially perform specific job without significant hazard	A. Any inquiry into past or current medical conditions not related to position applied for B. Any inquiry into workers' compensation or similar claims
10. Citizenship	A. Whether a U.S. citizen B. If not, whether applicant intends to become one C. If U.S. residence is legal D. If spouse is citizen E. Require proof of citizenship after being hired F. Any other requirement mandated by the Immigration Reform and Control Act of 1986, as amended	A. If native-born or naturalized B. Proof of citizenship before hiring C. Whether parents or spouse are native born or naturalized
11. Photographs	May be required after hiring for identification purposes	Require photograph before hiring
12. Arrests and convictions	Inquiries into conviction of specific crimes related to qualifications for the job applied for	Any inquiry which would reveal arrests without convictions
13. Education	A. Inquiry into nature and extent of academic, professional or vocational training B. Inquiry into language skills, such as reading and writing of foreign languages if job related	A. Any inquiry which would reveal the nationality or religious affiliation of a school B. Inquiry as to what mother tongue is or how foreign language ability was acquired

TABLE 5-1 The Employment Interview Guidelines (continued)

Inquiries Before Hiring	*Lawful*	*Unlawful**
14. Relatives	Inquiry into name, relationship and address of person to be notified in case of emergency	Any inquiry about a relative which would be unlawful if made about the applicant
15. Organizations	Inquiry into memberships in professional organizations and offices held, excluding any organization, the name or character of which indicates the race, color, religion, sex, national origin, handicap, age or ancestry of its members	Inquiry into all clubs and organizations where membership is held
16. Military service	A. Inquiry into service in U.S. armed forces when such service is a qualification for the job B. Require military discharge certificate after being hired	A. Inquiry into military service in armed service of any country but U.S. B. Request military service records C. Inquiry into type of discharge
17. Work schedule	Inquiry into willingness to work required work schedule	Any inquiry into willingness to work any particular religious holiday
18. Other	Any question required to reveal qualifications for the job applied for	Any non-job related inquiry which may elicit or attempt to elicit any information concerning race, color, religion, sex, national origin, handicap, age or ancestry of any applicant for employment or membership

TABLE 5-1 The Employment Interview Guidelines (continued)

Inquiries Before Hiring	*Lawful*	*Unlawful**
19. References	General personal and work references not relating to race, color, religion, sex, national origin, handicap, age or ancestry	Request references specifically from clergymen or any other persons who might reflect race, color, religion, sex, national origin, handicap, age or ancestry of applicant

I. Employers acting under bona fide affirmative action programs or acting under orders of equal employment law enforcement agencies of federal, state, or local governments may make some of the prohibited inquiries listed above to the extent that these inquiries are required by such programs or orders.

II. Employers having federal defense contracts are exempt to the extent that otherwise prohibited inquiries are required by federal law for security purposes.

III. Any inquiry is prohibited which, although not specifically listed above, elicits information as to, or which is not job related and may be used to discriminate on the basis of, race, color, religion, sex, national origin, handicap, age of ancestry in violation of law.

*Unless bona fide occupational qualification is certified in advance by the Ohio Civil Rights Commission.

Source: "Questioning Applicants for Employment: A Guide for Application Forms and Interviews Under the Ohio Fair Employment Practices Act." Distributed by the Ohio Civil Rights Commission.

The interview can also be used to clarify and expand on information presented in the written application. This is especially helpful when the applications are standard forms that allow little space for presenting more than the most rudimentary information.

Notes should be taken during the interviews for future reference, as it is often difficult to remember which candidate said what after a few days have passed. Notes made during and after the interview will keep this clear. The interviewing officer should write a summary of impressions and central pieces of information that indicate competencies directly after each interview. Thus, when scheduling a series of interviews, it is important to schedule, and safeguard, 15 or 20 minutes at the conclusion of each interview in order to record the observations. These notes should be maintained for a reasonable length of time in case the search is extended or there is a challenge to the process or selection.

ACCOMPLISHMENTS VERSUS QUALIFICATIONS

When hiring for fire fighter and junior-level officer positions, emphasis is often placed on qualifications and experience. When hiring senior executive fire officers, however, the focus should be shifted to accomplishments. The fire service wants leaders in its topmost positions. Leaders have track records of having accomplished projects, of having made changes.

Many people have qualifications to do a job, but some of these individuals simply perform to the minimum required standards. Others have not only the qualifications but the motivation to use these qualifications to improve the operation of a fire department. These are the individuals who should occupy senior officer positions.

Thus, the qualifications component of a job description for an assistant chief in charge of EMS might say "4 years experience as shift officer in charge of EMS." The application or interview should have one section that focuses on accomplishments during those 4 years. One EMS captain may have run the EMS system without any trouble (business as usual). An EMS lieutenant may have secured external funding for an additional squad, instituted a preventative public health program, and reduced the number of false calls by 20 percent. In this situation, the EMS captain has a higher rank and might therefore be considered more qualified by this criterion. However, the EMS lieutenant actually accomplished more though a lower-ranking officer. Leadership qualities are demonstrated by past accomplishments. These accomplishments can also indicate future potential. These issues need consideration when choosing between candidates.

REFERENCES

Checking references is one way to identify the strengths and weaknesses of candidates. These references should not only include superiors for whom the officer worked, but also peers and, especially, subordinate officers and fire fighters who worked for the candidate. Supervising officers sometimes have a very different view of an individual's capabilities and activities than do subordinates. It is important for a fire department to know how a new officer is likely to work with fire fighters and officers under his or her command. This does not necessarily mean that an officer who receives a harsh review from fire fighters should be disqualified. Some fire officers are not liked by the fire fighters, but they do a good job and are fair and equitable—even though they are also demanding and firm. In situations where the officer is likely to encounter tough discipline problems, this may be a desirable trait.

When contacting references, it is important to preserve confidentiality. Information gathered should not be shared with anyone outside of the hiring process. Decisions and comparisons among the candidates should remain in the confines of the group active in the process. The candidates' strengths and weaknesses should not be a topic of conversation outside the selection process. This can stereotype incoming individuals and negatively impact their ability to perform their duties.

The open records laws require that unless a candidate waives his or her right of access in writing, he or she has the right to view any written references that were submitted. Fire departments must account for this situation by providing waiver forms to candidates. In addition, it is important to let references know that their letters must be made available to the candidate on demand.

INTERNAL VERSUS EXTERNAL CANDIDATES

The question of whether to hire candidates from within or outside the department is a debate with no clear answer. It depends on the department and the situation at hand. Each alternative can offer a number of benefits and each has a number of drawbacks as well.

Some of the positive aspects of an internal candidate include: familiarity with the department, knowledge of the personnel, knowledge of the internal politics, knowledge of the existing contracts, and familiarity with the community. Some of the strengths of an outside candidate include: ability to introduce new ideas, opportunity for reexamination of the department, establishment of new personal interactions, and changes in the existing political power structure.

These attributes need to be considered along with candidate qualifications when making the final selections for executive officer positions. Remember, there is no consistently right answer to whether it is best to hire from within or outside the department.

SUMMARY

Hiring provides an opportunity to reexamine the operations of the fire department in order to determine what qualifications and characteristics of new personnel would most benefit the department. It is a form of department renewal that can motivate improvements in morale, operations efficiency, and community service effectiveness.

Hiring is a challenging process that requires adherence to legal mandates. In order to hire the most appropriate person for the job, it is first necessary to have an accurate description of that job. Once this is completed via a job analysis, a comprehensive, results-oriented job description can be written. This job description is then understood uniformly by all individuals concerned. It can also be utilized for promotions, job expansion, or disciplinary actions, as necessary.

A number of recruitment outreach techniques are effective in different communities, or when trying to recruit from different segments of the community. Additionally, volunteer and part-paid departments may want to utilize different recruiting techniques than do career departments. The recruitment process, just like the hiring process, must be customized to fit the characteristics of the community and the needs of the fire department.

It is important to follow appropriate screening and testing procedures once the applicants are identified. All appropriate legislation must be followed and documented carefully. Testing must meet validation criteria, and the process must be objective. The type or use of a test is dependent not only on local regulation, but also on the type of position and the qualifications of the candidate being solicited. Searches for executive fire officers should identify accomplishments and match capabilities and orientation to department needs and the organization's culture. Thus, a variety of techniques are considered according to the position and applicants involved.

Hiring is a critical activity. It determines the staffing composition for years to come. This process must be carefully followed and monitored to ensure that the best possible individuals are hired.

CASE STUDIES

Case #1

Chief Marks has utilized testing for incoming fire fighters in the past. He has recently secured permission to add more part-time fire fighter/paramedics to the department's staff. He asks his EMS assistant chief to help him compose a test.

The chief selects questions from a number of nationally recognized training manuals. The assistant chief reviews the quizzes that were used at the end of continuing education sessions conducted by the medical advisor and selects one question from each quiz given over the last 18 months.

1. What is wrong with this test?
2. How would you try to validate this test?
3. How should the test be constructed?

Case #2

A fire department with a full-time chief and no other full-time employees has an assistant chief, two captains, and three lieutenants. Promotions to lieutenant have always been based on test scores, but promotions to captain and assistant chief have been based on seniority in rank. The officer roster is as follows (the names in rank are listed in descending order of seniority):

Chief Wesleyan
Assistant Chief Davis
Captain Raddich
Captain Varney
Lieutenant Murray
Lieutenant Mendoza
Lieutenant Jones

Assistant Chief Davis resigns, a lieutenant's test is given, promotion is based on seniority, and the roster is changed to reflect the following:

Chief Wesleyan
Assistant Chief Raddich
Captain Varney
Captain Murray
Lieutenant Mendoza
Lieutenant Jones
Lieutenant Wallace

Captain Murray resigns. The chief decides to have a promotional test for the open captain's position. All lieutenants are eligible to take the test. Lieutenant Jones receives the highest score and is promoted to captain. The chief and assistant chief constructed the test by making up questions and taking questions from self-study texts.

1. What is wrong with this procedure?
2. What is wrong with the test?

The chiefs could argue that they were simply making the promotion process more objective and professional, except that when Assistant Chief Raddich retires a year later, Captain Varney is promoted to assistant chief on the basis of seniority and Lieutenant Mendoza is promoted to captain on the basis of seniority.

3. Does the chief have any valid justification for his change in practice?
4. Would it be appropriate if the chief promoted for the position of assistant chief on the basis of seniority and then tested for captain again?

Case #3

Over the past 7 or 8 years, the Full-time Fire Department has experienced a steady increase in the number of EMS calls received. Part of the growth is attributed to several recent annexations, along with the fact that the community's population is aging.

Recognizing that the current fire department's organizational structure and physical plant facilities could not be enlarged to accommodate the additional emergency response service demands, Chief Sanchez petitioned the city council 2 years ago to approve the construction of a new station on the west side of town. His request was approved, and the new station construction project is nearing completion. Since the project began, four fire officers have chosen to retire, leaving the department with a number of vacancies in the officer ranks. No promotional exam has been scheduled since the chief intends to post the test and hire replacements for the retired officers when he appoints additional officers for the new west side station. The chief says he has been reluctant to post the exam until close to the opening of the new station because of severe budget restrictions. Now that the station opening is fast approaching, a number of fire department members are asking for the posting date so that they can begin to prepare. The situation is causing some tension among members of the department.

Although Chief Sanchez knows that action must soon be taken, he continues to delay such action until he can assess the role of station officers in his department in terms of the needed level of technical capability and the team player skills required. While most department members are technically well prepared to assume the role of station officer, only a very few are known to be team players who will not "rock the boat."

In order to select those officers the chief has identified as being technically prepared and who will accept control from the top without question, a written exam is developed by the fire department testing board with assistance from local educators. Within 2 weeks after the test is completed, the chief posts the promotional notification. The only qualifications identified for becoming a candidate for promotion to officer rank are: (1) being a certified paramedic and (2) having served on the department for 5 years prior to the posting date. In addition to the written exam, all candidates are required to be interviewed by both the senior fire officers and the fire board.

Following certification of the test results by the fire department testing board, the chief is somewhat dismayed because three of the top five candidates are not those identified by Sanchez as being "acceptable" candidates. After considerable deliberation, the chief decides to request that the supervisors of the top six candidates prepare a comprehensive evaluation of each candidate, covering a period of the last 5 years. In addition, Sanchez orders that a psychological evaluation be done on the six candidates.

1. Is it in the best interest of the fire department to promote only those officers who will not "rock the boat"?
2. Is the 5-year written evaluation by supervisors an appropriate measure of performance and capability?
3. Is the sudden additional requirement of a psychological evaluation appropriate?
4. How should the candidates respond?
5. As a member of the fire board, what would you do?

ENDNOTES

1. David J. Cherrington, *Personnel Management: The Management of Human Resources,* 2nd ed. (Dubuque, IA: Wm. C. Brown Publishers), p. 119.
2. Ernest J. McCormick, "A Study of Job Characteristics and Job Dimensions as Based on the Position Analysis Questionnaire," in *Journal of Applied Psychology,* August 1972, pp. 347–368.
3. Donald E. Klingner and John Nalbandian, *Public Personnel Management: Context and Strategies,* 2nd ed. (Englewood, NJ: Prentice-Hall, 1985), pp. 155–158.
4. Groups have organized to promote women and minorities in the fire service. Two examples are Women in the Fire Service (contact Terese Floren, P.O. Box 5446, Madison, WI 53705) and the United Black Fire Fighters of Akron, OH.
5. Copies of Title 7 of the 1964 Civil Rights Act and the 1972 EEO Act can be ordered from the U.S. Government Printing Office, Washington, DC 20402-9328.

REFERENCES

Ammons, David N., and James J. Glass. 1989. *Recruiting Local Government Executives: Practical Insights for Hiring Authorities and Candidates.* San Francisco: Jossey-Bass.

Bell, Arthur N. 1992. *Extraviewing: Innovative Ways to Hire the Best.* Homewood, IL: Business One Irwin.

Cherrington, David J. 1987. *Personnel Management: The Management of Human Resources.* 2nd ed. Dubuque, IA: Wm. C. Brown Publishers.

Connolly, Kathleen Groll. 1991. *Competing for Employees: Proven Marketing Strategies for Hiring and Keeping Exceptional People.* Lexington, MA: Lexington Books.

Decker, Kurt H., and H. Thomas Felix, II. 1991. *Drafting and Revising Employment Handbooks.* New York: Wiley Law Publishers.

Klingner, Donald E., and John Nalbandian. 1985. *Public Personnel Management: Context and Strategies.* 2nd ed. Englewood, NJ: Prentice-Hall. pp. 155–158.

McCormick, Ernest J. 1972. "A Study of Job Characteristics and Job Dimensions as Based on the Position Analysis Questionnaire." *Journal of Applied Psychology.* August, pp. 347–368.

NFPA 1021, *Standard for Fire Officer Professional Qualifications,* 1992 ed. Quincy, MA: National Fire Protection Association.

Pinsker, Richard J. 1991. *Hiring Winners: Profile, Interview, Evaluate: A Three-Step Formula for Success.* New York: American Management Association.

Weitzel, James B. 1992. *Evaluating Interpersonal Skills in a Job Interview.* New York: Quorum Books.

CHAPTER 6

Motivation and Discipline

Motivation, discipline, and mentoring are administrative tools that promote high-quality performances from employees. This chapter addresses aspects of motivation and mentoring, then deals with discipline.

Motivation and discipline are two sides of the same coin. The chapter begins with a general discussion of motivation. Motivation provides positive encouragement, recognition, and rewards so that members of the department will accomplish their tasks with enthusiasm and responsibility. Theories of motivation are presented along with a guide to applying these theories in the fire service. The importance of motivating superiors as well as subordinates is discussed, followed by a presentation on mentoring.

Mentoring is a special professional relationship administrators can develop to facilitate growth in specific administrative fields. It provides detailed and concrete assistance as well as personalized interaction with an expert in a specific area. Mentoring is receiving considerable attention in the business world as research continues to identify ways mentoring has aided the development of mid- and upper-level managers. Recently the public sector has initiated mentoring programs to help mid-level managers improve their capabilities.

Employees who fail to accomplish their assigned tasks or disobey department rules or regulations must be disciplined. Discipline rectifies inappropriate action in order to produce proper employee activity. The discussion focuses on positive discipline, and potential applications to the fire service are outlined.

Both motivation and discipline are important tools for fire officers in their daily interactions with subordinates. This chapter provides a good grounding in primary motivation and disciplinary approaches that can be applied to the fire service.

MOTIVATING EMPLOYEES

Motivation can be defined as providing an inner urge that prompts a person to action with a sense of purpose. This positive sense of purpose and the resulting action result in a different quality of performance on the part of fire fighters and other employees. Human resource management research has shown a correlation between motivation and job performance. Motivation develops an attitude that makes an individual desire to be productive and do a job well. This leads to improved performance, increasing the efficiency and effectiveness of an organization. In fire departments motivation improves morale. Fire fighters with high morale are innovative and creative. They like challenges and work hard to think of ways to surpass obstacles and perform the necessary work at a higher and more professional level.

Motivation improves the function and operation of a fire department and results in better service to the residents of the community. Improvements may take the form of higher-quality performance, more rapid response, or increased professionalism. Motivation is a valuable leadership skill that can produce visible results.

Indications of High Morale and Motivation

Motivated fire fighters can be identified in a crowd. These are the fire fighters with pressed, crisp uniforms, polished shoes, and neat personal appearance. Their stations are usually well maintained, and the apparatus and equipment are always ready for the next response. Fire fighters with high motivation take pride in their station and want it to reflect their commitment to the fire service.

Volunteer and paid personnel who are satisfied with their work will enjoy coming to work. Maintaining a high level of job performance requires motivation and dedication regardless of paid or nonpaid status. Unmotivated volunteer fire fighters often stop participating in depart-

FIGURE 6-1 Fire fighters with high levels of motivation take pride in the appearance of their apparatus.

ment activities, potentially limiting the department's ability to deliver service to the community. Thus, motivation in volunteer fire departments is as important, or even more critical, than in career departments.

Challenges to Motivation

If motivation yields positive outcomes, why aren't the frameworks for motivating employees established by all administrators? Even under the best of circumstances, there are conditions challenging efforts to motivate. Some of the best efforts to stimulate personnel have been thwarted by tradition, policies, rules, regulations, misunderstandings, poor communication, and cultural differences. As a result, in any concerted effort to motivate employees, the fire officer must consider the nature of the organization and the limit to his or her discretionary power.

Motivation poses the greatest challenge to leadership. It is often the most difficult part of a fire officer's job.

Labor Agreements

There are a number of motivating techniques presented in the next section of the chapter, but fire officers may be limited in their choices. In fire departments operating under a labor agreement, fire officers are restricted to motivation techniques that meet the guidelines of the agreement.

One of the most traditional motivation techniques is paying a bonus to employees who demonstrate outstanding service. In fire departments where compensation is fixed by contract, however, such bonuses cannot be given, a circumstance found in most public personnel systems.[1] A second motivational tool is using promotional opportunities or creating special assignments to reward outstanding employees. This option is not appropriate because promotional opportunities and special assignments are developed in accordance with civil service and labor agreements. Special perks such as being allowed to take a half day off are also restricted.

These restrictions, while limiting techniques to motivate, also provide equal opportunity for all members of the fire department. Labor agreements and civil service restrictions limit decisions based on personal preference and personality as opposed to competency, ability, and effort. In some instances these limits can be detrimental, but they can provide objective measures to qualify rewards.

Current research identifies positive and negative aspects of motivation in the public sector. The Volker Commission on Public Service is a blue-ribbon panel of prominent citizens from the public and private sectors charged with identifying ways to enhance public service as a career. Commission and related research identifies a lack of motivation among senior administrators. In addition, a number of good executives are leaving public service. Much of this is due to lack of motivation and demoralization resulting from political manipulation.[2] Another study dealing with new employees at all levels of the public sector concluded that "public sector employees manifest significantly higher levels of job satisfaction than their private sector counterparts."[3] Other studies of both large and small groups of public employees also point out that job satisfaction is equal to or greater than that in private industry.[4]

Challenges to Motivating Volunteers

Fire officers also face challenges to motivating members of the volunteer segment of the fire service. Budget restrictions often limit the ability of fire officers to send department members to special off-site training, which can be an excellent motivational experience. However, many volunteer fire departments simply cannot afford to send fire fighters.

The limited amount of time that volunteers spend "on station" restricts opportunities to motivate these individuals. Fire officers have to create opportunities to let fire fighters know they are valued. It is important for fire fighters to know their contributions to department operations are appreciated.

Some officers assume that because volunteer fire fighters donate their services they are automatically motivated. This is only partially true. It has been said, "Officers should *ask* a volunteer to do something and *tell* a career fire fighter to do something." The assumption here is that by asking, the officer is acknowledging the volunteer's contribution. This is a form of recognition, and should be motivating. On the other hand, career personnel are being paid for their services and, thus, can be *directed* to perform tasks. There is fallacy in both parts of this argument.

The idea that an officer should tell a career fire fighter to do something is based on the assumption that a career fire fighter is not motivated and does not need to be motivated. Both of these assumptions are false. In addition, asking a volunteer assumes all volunteers are motivated and the only further motivation necessary is to be asked rather than told. These assumptions are also false. Volunteer fire fighters can become burned out and develop low morale.

Motivation must be practiced in volunteer as well as career departments. *All* employees desire recognition, the feeling of being valued, and the feeling of being a part of the organization. Most of the theories and techniques of motivation presented later in this chapter can be applied to volunteer, paid on-call, combination, and full-time fire departments. Although the restrictions to motivation may differ in various types of fire departments, the types of motivation techniques that are limited, such as pay bonus, and the types that can be used, such as public recognition, tend to be the same. Thus, although some departments face greater challenges to motivating volunteers, the techniques are similar and the need is just as great.

Motivation Technique	*Potential Restriction*
1. Pay increase bonus	Labor contract
	Civil service
2. Promotion	Civil service
	Labor contract
3. Assignment	Labor contract
	Civil service
	Staffing
4. Recognition	Time
	Schedule
5. Praise	Time
	Schedule

The first three techniques are restrictions the fire officer cannot control. Recognition and praise are restricted by time and schedule, which can be structured by fire officers. Fire officers must *make* time and arrange the schedule to recognize and praise fire fighters. The resulting increase in motivation is worth the effort.

Effort on the Part of Leaders

Despite the limits on fire officers' discretionary powers, motivating fire fighters and officers is one of the most important aspects of administration and must be recognized at all levels. At a minimum, a fire officer can do two things to motivate: (1) notice the actions of subordinate members and (2) try to motivate.

Fire officers should notice *all* actions of their subordinates. Dwelling only on the negative draws attention away from positive actions. Discipline and praise are both important. If one receives only negative feedback on activities, there is no motivation to do an outstanding job, only to do just well enough to keep from being noticed. This is highly

FIGURE 6-2 Recognition for good performance, either formal or informal, is an important way to motivate fire fighters.

detrimental. When fire fighters have answered an unusually stressful call and performed as they have been trained to do, it is important to say, "Nice job, folks." Acknowledging that they performed their jobs under adverse conditions in a stressful environment is often more valuable than any physical or monetary reward.

Fire officers have a responsibility to *try* to motivate. Even if the avenues to visibly recognize staff are limited, the fact that an officer admits this and tells the fire fighter that he or she deserves a reward that is not coming validates an individual's contribution. Verbal appreciation and acknowledgment are valuable motivational tools. Fire fighters know when an officer is making a genuine attempt to motivate and recognize performance, and they appreciate the effort.

Motivational Theories and Techniques

Fire officers should be creative in their efforts to lead and motivate employees. Just as each department is different, each department member is different. As a result, officers must understand a variety of motivation techniques and learn how to use them effectively. The situational theory of motivation states that what works in one situation will not work in another. If one approach isn't working, it may be necessary to modify the technique or switch to another. Some of the most current literature explores the debate over which motivational technique is most appropriate. The result? "The truth of the matter is that, while light has been shed on some of the most obscure aspects of the motivational process, few irrefutable answers have been found; some evidence supports each of the theories of motivation. . . . (T)he safest assumption is that no one theory is adequate to explain a phenomenon as complex as employee motivation."[5]

Most techniques practiced today were based on research into employee motivation. A number of classic theories are still applied because various studies continue to support their validity. Specific techniques for putting these theories into practice have also been developed.

Theory X and Theory Y

One classic theory of motivation, known as Theory X and Theory Y, was expressed by Douglas McGregor.[6] McGregor identified two belief systems about the nature of workers, each composed of a different set of perceptions about employees, that determine what type of motivation technique and leadership style will be most effective.

Theory X

- Workers dislike work and will avoid it if possible.
- Workers must be coerced, directed, threatened, and controlled.
- Workers desire direction, avoid responsibility, have little ambition, and want security.

Theory Y

- Work is as natural as play and rest.
- External control and threats are not the only means of getting people to work.
- In appropriate circumstances, employees learn to accept and seek responsibility.
- Under authoritarian management, the abilities of employees are only partially utilized.

These two ways of viewing employee characteristics have direct implications for how management would try to motivate. Managers who subscribe to Theory X believe that only money and time are appropriate motivation techniques and that even those will not make a difference in performance. Managers who subscribe to Theory Y believe otherwise. These managers realize that people who want to work will do the best they can, and recognition and reward tend to enhance their efforts.

Theory Y is based on a different understanding of employees and how they can be motivated to do more. It was the first theory to recognize the inherent desire of certain employees to work and make a contribution to the organization.

Maslow's Hierarchy of Needs

One of the most well-known theories that attempts to explain why employees respond differently to different motivations is Maslow's hierarchy of needs.[7] Maslow's hierarchy is often termed a "need-based" motivation technique because it relies on satisfying a need as a means of motivation. Maslow maintained that each individual has five levels of need. The fulfillment of each of these needs is a motivation factor. However, once a lower-level need is fulfilled, the employee is motivated only by something that addresses the next highest need. The chart below lists the five levels identified by Maslow and how they translate into potential motivation on the fire department. For example, the basic physical need can be met through a salary that enables a fire fighter to purchase a home and feed a family.

Five Levels of Need

1. Physical—food, shelter/salary, and benefits
2. Safety, security—job security/work safety practices
3. Acceptance by others, love—recognition of work performance/acceptance by family
4. Self-esteem—feeling good about oneself/proud of the job one does
5. Self-actualization—personal development

These areas of need can be affected both on the job and in personal life. Acceptance and love deal with the human need to be accepted by other people, especially people who matter to the individual. A fire fighter needs acceptance from other fire fighters and support and acceptance from his or her family. For example, if a fire fighter is experiencing marital difficulties, he or she is facing rejection in a personal part of life. This can have a negative impact on the individual's job performance—he or she may lose interest in the job, be unable to focus on tasks, and lose motivation. Sometimes a demonstration of acceptance on the job can mitigate the impact of personal difficulties and motivate the fire fighter. This can be done through formal employee assistance programs, one-on-one discussions, or even informal conversations with other fire fighters.

The highest need level, self-actualization, is often the most difficult to understand and to use as a motivational tool. People achieve personal development at various career levels; not every fire fighter wants to be an officer. There are those who want not only to be a pump operator but also the *best* pump operator. Self-actualization means feeling that one is developing personally and has the opportunity to continue to do so until reaching the performance level desired. This underscores the need for personal development programs in fire service training and education.

Expectancy Theory

The expectancy theory, developed by Victor Vroom, is based on the concept that "individuals function rationally and calculate the value of actions."[8] Expectancy theory simply states that people are motivated to work if they believe their efforts will be rewarded and they value the rewards offered.[9] In order to determine whether a motivation technique will work, fire officers must answer the following three questions from the perspective of the fire fighter:

1. How much is the task worth? (What is the reward?)
2. What will it take to do it? (How much effort will the task take?)
3. Can I be successful? (What are the probabilities of accomplishing the task?)

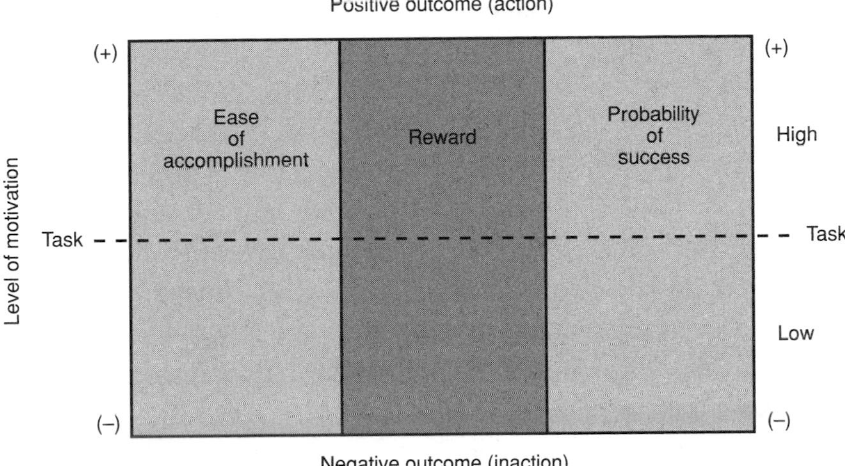

FIGURE 6-3 Expectancy theory states that motivation is based on three elements that, when added together, must result in a positive outcome to encourage rational people to take action.

If the reward is not equal to the task, or the effort is greater than the reward or importance of outcome, fire fighters will not be motivated to undertake the task. If the chance of success is small, the willingness to approach the task is also small; no one likes to fail. Thus, fire fighters make rational choices concerning when and where to put forth maximum effort. Answering these questions can help fire officers set rewards and time expectations for task performance to guarantee that the expected outcome encourages the fire fighter to perform the task required.

A comparable motivational theory is Livingston's Pygmalion theory (also known as the self-fulfilling prophecy),[10] which maintains that subordinates often perform according to their manager's expectations. High expectations from fire officers result in high quality and high productivity by fire fighters, whereas low expectations result in low quality and low productivity.

This can be seen in departments where officers continuously brag about the performance of their fire fighters versus those departments where officers berate a certain shift. After enough put-downs, the members of the shift are not motivated to try to be anything more than the officer said they were. In contrast, departments with high expectations for performance often surpass the stated expectations of the officers and members.

Motivation-Hygiene Theory

Herzberg identified only two elements of motivation: hygiene needs and motivation needs. Hygiene needs relate to an individual's dissatisfaction in the workplace. If met, they have little effect on an employee's satisfaction or dissatisfaction; however, if they are not met, they will contribute significantly to employee dissatisfaction. Motivation needs, on the other hand, are those that encourage individual growth and job satisfaction. The motivators identified by Herzberg have proved to promote a higher level of job satisfaction among employees in the public sector.[11]

Hygiene Needs

Work conditions
Supervisory relationships
Salary
Administrative Policies

Motivation Needs

Job advancement
Additional responsibility
Intrinsic satisfaction from the job
Recognition of one's efforts
Personal achievement

Goal-Setting Theory

The goal-setting theory, developed by Craig Pinder, states that human behavior is based on goals and objectives, and higher goals will result in higher performance levels. Clear and measurable goals are better than ambiguous ones so that fire fighters know exactly what is expected of them. Ambiguous goals result in frustration that affects job performance. Incentives will not work unless they result in clear and ambitious goals. (This aspect of goal setting is developed further in Chapter 10, Planning, which discusses individual planning for personal development as well as fire department organizational planning.)

An example of a positively motivating goal is to ask an engine company to preplan an industrial site within 2 weeks of participating in a 4-hour training course, using the results of a fire inspection, blueprints, and site maps. The fire fighters have the training, information, and time necessary to accomplish the goal. Asking the company to preplan a facility without providing any information would be unrealistic and frustrating.

Applying the Motivation Theories

Fire officers can apply the basic tenets of these theories in a variety of ways. Some of the motivation elements such as pay and benefits cannot be controlled by an individual officer. However, a number of other work conditions can be affected. These can range from keeping a neat station to formalizing a series of personal development conferences with each subordinate. Other examples of motivation techniques include:

- Instilling a sense of pride in the department
- Demonstrating pride in the performance of fire fighters
- Maintaining the station and apparatus in a manner that displays pride to the public
- Involving fire fighter/paramedics (and other subordinates) in decisions and planning
- Communicating openly with department members (secrets breed rumors, which create distrust and erode loyalty)
- Publicly recognizing department member achievements
- Demonstrating interest in department members' personal growth
- Recognizing fire fighters' and junior officers' attempts and desires to contribute to the growth and development of the department
- Using special talents and skills of members—regardless of rank
- Listening—and responding positively and honestly
- Responding to new ideas—not discarding them without positive consideration
- Not depending on apparatus and equipment acquisitions as primary motivators

These are some of the many ways to implement the concepts behind the theories. If fire officers understand the theories behind motivation techniques, they can adapt those techniques to fit their individual departments and circumstances.

Motivation Throughout the Organization

Most discussions about motivation focus on motivating subordinates, downward in the organization. However, superiors also need motivation. Officers with low morale suppress innovative ideas and discourage initiative. It is sometimes left to midlevel officers, or even fire fighters, to motivate senior officers to get the support they need. Recognizing the efforts of the officers and the difficulties they encounter in managing the organization will boost morale and encourage high-level performance.

Case Study #1 at the end of this chapter demonstrates how a soured attempt to do something positive severely curtailed a senior officer's motivation to make changes. As you consider this case, recognize that the fire chief has probably lost much of his motivation to do something for the department members. In this instance, if fire fighters in the department recognize the chief's effort to make a change, the chief may rethink his plan of action. Even if individual fire fighters were to let the chief know that they appreciate his good intentions, they would provide the motivation necessary to keep trying.

Mentoring

Mentoring can be viewed as a motivating process. The individual serving as the mentor, sharing experience and expertise, is recognized for abilities and knowledge and has the opportunity to see the impact of his or her involvement on another individual. The protégé, or individual in the learning role, feels important and worthy enough to be given individual consideration and assistance. Both parties in the process generally have high motivation levels as a result of the experience.

Mentoring is more than just a motivating technique, but its contribution to morale and motivation warrant discussion at this point.

In order to understand and implement a mentoring program, a more detailed discussion is necessary.

What Is Mentoring?

Mentoring is an activity that pairs experienced officers with new or inexperienced officers to facilitate their career growth, enhance officer performance, and stimulate a greater understanding of the organization in addition to teaching leadership and administrative skills. The role of the mentor is to motivate and encourage the protégé to reach full potential. In order to do this, the mentor must create an open atmosphere that promotes trust and confidence. The mentor offers advice, responds to issues and problems the protégé encounters, and acts as a role model. Much of the exchange between mentor and protégé must be treated with confidentiality. Exchanges of information, ideas, and opinions are not meant to be shared beyond the relationship without the consent of the other party. How to select the right mentor to fill this role will be discussed later.

Mentoring relationships benefit both the individual and the fire department. They can be a method of support by which interaction between two individuals (mentor and protégé) cultivates the knowledge and skills necessary to manage human interactions within the fire department. Mentoring also helps the protégé acquire a better understanding of the organization's structure, culture, and procedures. It is a learning process that facilitates improvement of the protégé's self-confidence, enhances promotional opportunities, and teaches the individual to cope with the formal and informal structure of the organization.[12]

Mentoring will cultivate leadership skills, strengthen one's ability to take risks, provide operational direction, and inform the protégé about what's going on. The process helps the protégé understand that team building and cohesiveness are effective tools in creating a work environment that is responsive to the needs of the community and the fire department; the very process of mentoring serves as a model of how administrators can work cooperatively.

Unlike traditional fire service training programs, mentoring activities are not rigidly structured. Quite often, officer training programs are organized into various formats and topics and use a variety of learning techniques and styles. When adult learners desire to improve their knowledge and skills, the most widely accepted approach is to pursue higher education or enroll in some type of organized training or self-study program. These traditional programs are highly structured and require specific commitments of individual time and resources. The process of mentoring, on the other hand, is an ongoing activity that has no time constraints or physical boundaries.

The Responsibility of Senior Officers in Developing Junior Officers

Senior fire officers have a vested interest in the careers of junior officers and prospective officer candidates. As the fire service transcends the traditional model toward one that responds to the constantly changing needs of the community, so too must the department's leadership. The question is how can this be accomplished given the paramilitary structure of the organization and the inherent staffing restraints placed on it by such a system? One effective method is a mentoring program.

Traditional officer training and development programs often function within the parameters of the student–teacher (instructor) relationship. Junior officers and prospective candidates participate in a series of scheduled learning sessions, usually conducted in a classroom setting. Instructors and senior officers present the necessary information, which junior officers are expected to absorb and translate to their particular situation. Unfortunately, insight into how their own fire department actually operates under emergency and nonemergency conditions is missing. Mentor–protégé (senior–junior) relationships help fill the gap between the technical competencies taught in the classroom and the administrative skills and knowledge a fire officer needs to understand "the system" and make a difference in the final outcome.

Establishing a Mentoring Relationship

The selection of a mentor deserves considerable deliberation since this is the person who will help you build the foundation upon which your personal and professional growth will occur. Choose your mentors carefully! A prospective mentor should express a commitment to the process and demonstrate a level of competence that is compatible with your needs. Without this commitment to you, the activity is reduced from a mentor–protégé relationship to one of an advisor who responds to questions from the protégé only when called upon.

The competence factor is equally important since you will rely on this person to provide guidance and counsel in affairs related to the advancement of your career. If the mentor is not competent in the areas you have identified as weak, the end result will be disappointment and frustration. A word of caution: Don't select a mentor who demonstrates only technical competence. Expert power in a technical field such as the fire service is no guarantee that the person you select will be able to function as a counselor or teacher in matters concerning employee motivation, communication skills, organizational politics, or other dimensions of administration that may surface.

Another interesting and important aspect of the mentor selection procedure is the influence of one's personality on the relationship. Be certain that you are compatible with the prospective mentor and feel comfortable with him or her. Be cautious about entering into a relationship that is bound by the charismatic style of the mentor if you are an analyzer or supporter. (See Chapter 2, Leadership.) Conversely, if you are a gregarious person who feels comfortable with director or persuader types, you may not want a mentor who is apt to stress technical knowledge and skill development over human relations issues. The conflict of styles may force you to spend more effort relating to the mentor and detract from gaining the knowledge associated with the mentoring experience.

As you start to identify prospective mentors, make sure you have defined the goals and objectives you wish to accomplish by this process. Without clear and concise goals and corresponding specific objectives, the task of identifying the "right person" will become cumbersome and frustrating.

Begin the search by talking to fire officers in other divisions of the department or contact officers from other departments. Remember, the selection of a mentor is not restricted to a person who holds a position superior to yours. Effective mentors can be officers who have

FIGURE 6-4 To be successful, a mentoring relationship requires effort and commitment on the part of both parties.

specific types of expert power or those who "just do a great job of managing."

Determine whether the mentor you select has the time to help you develop your skills. In most instances, the individual will be flattered at being asked and will respond positively. Explain your purpose for seeking a mentor–protégé relationship and present, *in writing,* your career goals. Clearly define the parameters for initiating the relationship and determine convenient times during which you can observe and interact with the mentor.[13]

The informal process may begin as a basic observation exercise whereby the protégé observes the mentor in a work setting and attempts to emulate his or her performance. Encourage the continuation of the relationship through regular conversations held at the fire station or even in a social setting. Even though the process is not structured, the protégé will begin to gain new knowledge that will contribute to career growth. At the same time, the protégé has developed a working relationship with the mentor that can be beneficial in the future.

Mentoring programs are becoming a point of increasing interest in the public sector. These programs match mentors and protégés, and provide a structured format for this relationship. Formal training to build mentoring programs has begun in a number of cities. Most of these are receiving careful scrutiny to measure their degree of success. The fire service, as well as the public sector in general, will be learning more about mentoring in the future. It is a positive way to change performance.

THE IMPORTANCE OF DISCIPLINE

Motivation and mentoring are both seen as positive methods of directing behavior. Sometimes the opposing management tool, discipline, is also required. Discipline, too, can result in a positive change in employee performance. As a result, it is an area in which fire officers must develop competence and comfort.

Discipline is used to maintain appropriate conduct in the work environment. There are different ways of viewing discipline procedures. Most fire officers and fire fighters see discipline as a punishment for doing something wrong. This is a negative view.

Discipline is also the way an officer can correct fire fighters' behavior to ensure they do not take the same action in the future. Discipline makes the individual a better employee and motivates other members of the department to work according to the rules and avoid the mistake their colleague has made. Discipline is not only a reprimand of someone who makes a mistake, but a reward to those who perform well. Applying disciplinary measures to a fire fighter who disobeys is an indirect way of recognizing those fire fighters who do things right.

Avoidance of Discipline

Most people find disciplinary action distasteful, whether they are receiving it or giving it. However, discipline is "essential to the success of modern managers;" it cannot be avoided.[14] Because close personal relationships are developed in fire departments, officers are often tempted to ignore minor disciplinary problems. This can have disastrous effects. When a fire fighter is not disciplined for an infraction of the rules, he or she quickly learns that this rule, and possibly others, are not important: there is no consequence for disobeying. This leads to flaunting authority and freelancing. Other fire fighters who have continued to obey the rules begin to feel discriminated against, questioning whether they would have been disciplined in the same situation. In addition, they have been shown that following the rules doesn't matter, and their motivation to continue to do so has been greatly decreased. Respect for rules—and for officers—begins to diminish. These departments develop low morale and low performance standards.

When discipline for rules violation is not uniform, when some individuals are allowed to flaunt the rules while others are not, morale and motivation decrease drastically. "Playing favorites" not only violates equal opportunity standards, it breeds disrespect for officers in authority and bad feelings toward fellow fire fighters. This can create a dangerous combination in emergency situations.

No one likes to be disciplined. However, when discipline is needed, it can serve more as a motivating factor than a punishing factor. Even those who receive disciplinary action often admit that it is appropriate. What is important is that the discipline be fair and just.

Approaches to Discipline

Discipline can be presented as a positive activity designed to improve the performance of the fire fighter and the overall operation of the department. Yelling at a fire fighter in front of other members of the department is not an appropriate form of discipline. Attacking an individual personally is totally inappropriate and has no place in formal disciplinary procedures. In fact, such action has no place in the fire service.

When taking disciplinary action, the fire officer must focus on the improper *actions*—not on the individual. The focus is corrective action, not belittling the individual. A fire officer who yells as a disciplinary technique, or makes comments referencing race, gender, ethnicity, or religion, is a candidate for disciplinary action.

Discipline motivated by a desire to punish rather than correct tends to have negative effects. Unfortunately, punitive approaches to discipline often motivate employees to achieve only the minimum acceptable performance standards. These fire fighters perform the exact duties required, but no more. This occurs in departments where punitive discipline is coupled with negative motivation or no attempts at motivation at all.

Constructive or positive discipline is an attempt to correct behavior and eliminate further errors on the part of the employee involved and others. It is done in a calm manner, and the disciplinary action taken matches the severity of the violation of conduct. Positive discipline requires the officer to employ swift, firm, and serious disciplinary action.

Constructive discipline is closely tied to a department's participation in employee development programs. When a department is committed to employee development, discipline is seen as a necessary and contributing component of that development process. It offers the opportunity to correct inappropriate behavior in a way that will reduce or eliminate that behavior by the fire fighter and, hopefully, by others as well. When problems are addressed in this manner, fire fighters tend to develop a strong sense of personal responsibility. When constructive discipline is used, morale improves and the number of disciplinary incidents decreases.

To have a positive approach to disciplinary problems, the department must have and follow a clear set of rules and expectations of conduct. Having rules that are not enforced is more damaging than having no rules at all because this teaches fire fighters that rules don't matter. These rules and regulations must be communicated to the fire fighters in understandable terms and format. They must then be followed by everyone—from the chief to the newest rookie. Often these rules include the appropriate procedure for taking disciplinary action. This tends to make discipline more uniform for the same infraction and establishes a sense of "equal justice" through the department.

Types of Disciplinary Problems

The seriousness of disciplinary problems may vary. Most can be placed in four categories: (1) rules violation, (2) unsatisfactory performance, (3) personal problems, or (4) illegal acts.

Rules violation is a failure to follow formal rules and procedures.

This includes the general department rules as well as SOPs. Although one would argue that failure to follow SOPs on an emergency response is more serious than failure to follow the rule of specified dress code, it must be understood that both deserve attention and discipline.

In departments with union agreements, the rules of conduct may be part of that agreement. In these instances a fire officer may have to deal with different interpretations of a rule and a potential grievance. This is why it is important for a department to have clear, understandable rules and procedures. If the department and the union are unable to come to an agreement, the grievance may go to arbitration.

Some rules, even if they are not formally written and adopted, have been upheld in court as appropriate expectation of conduct. These include:

- Insubordination
- Abusive language to a supervisor, subordinate, or fellow employee that is designed to embarrass, ridicule, or degrade
- Horseplay that causes a high risk of injury
- Fighting

Unsatisfactory performance occurs when a fire fighter simply fails to do assigned work. The work may take too long, include so many mistakes that it needs to be redone, or simply never be done. When this is due to lack of skills, the employee development and training program can be used to rectify the situation, and should not be considered disciplinary action. When the problem is motivation or unwillingness to perform, however, the fire officer should consider whether adequate and appropriate motivation has taken place and whether this can be coupled with disciplinary action to get the fire fighter to work.

Personal problems can interfere with work performance. In the fire department, personal problems that interfere with the safe and rapid performance of duty can endanger fellow fire fighters and the community. Therefore, such interference cannot be tolerated. If such behaviors are observed, fire officers may need to refer the fire fighter to an employee assistance program (see Chapter 3, Human Resources Management) or recommend counseling. It is important to remember that the fire officer is in charge of the safety of his or her fire fighters. Personal problems that interfere with safe operations cannot be tolerated.

Illegal or dishonest acts should never be tolerated. It is sometimes difficult to prosecute someone for an illegal act. However, like undisciplined misconduct, failure to prosecute, or at least discipline, signals approval of the action. This generally encourages the employee to continue to violate the law, and other employees may follow suit.

Progressive Discipline

There are a number of disciplinary actions that may be taken against an employee. Some are more severe than others. Generally, a fire officer starts with the least severe form of discipline and uses increasingly more severe measures for successive violations. However, the severity of discipline should correlate to the severity of the misconduct. This progressive system is based on the belief that employees have a moral as well as legal right to be told about their errors and to be given the chance to rectify them before adverse action is taken against them.[15] The following are the most common forms of disciplinary action. They are presented in order of increasing severity.

Verbal warning is a simple comment an officer makes to a fire fighter clearly stating that a certain act is not acceptable. This warning lets both the fire fighter and the officer know that the fire fighter now understands what is expected and what is not allowed. This eliminates the

FIGURE 6-5 Progressive discipline begins with the least severe approach and increases in severity with successive violations.

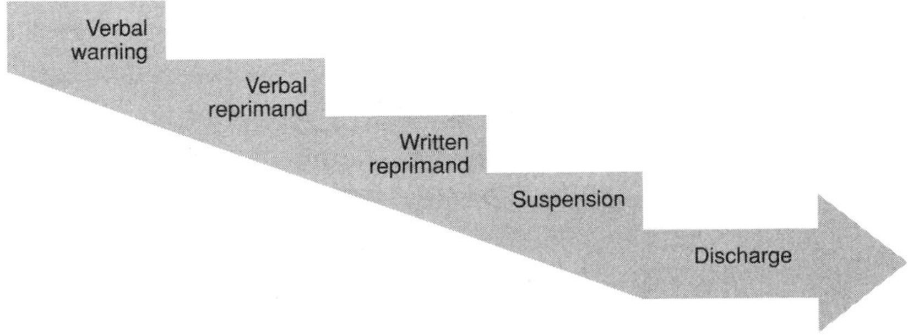

problem of misconduct due to ignorance. Although nothing is recorded in a verbal warning, the officer may refer to the incident during performance appraisal.

Verbal reprimand not only informs the fire fighter that the conduct is unacceptable, it also tells him or her that the activity must cease and performance must improve in this area. A fire officer should issue a verbal reprimand in private. The discussion must focus on the misconduct and on steps the fire fighter can take to improve. The fire officer should never belittle the fire fighter during this session: it should be a constructive session that clearly indicates that continuing misconduct will not be tolerated and explores alternative solutions. The fire officer should be certain the employee not only understands the problem, but also knows how to correct it. The fire officer should make a written note of the session in case further action becomes necessary.

Written reprimand is a written description of the problem and of the disciplinary action. Although it is similar to the verbal reprimand, this is a more formal procedure. After a verbal discussion with the fire fighter, the fire officer prepares a written record of what was said and decided. This becomes part of the fire fighter's personnel record. The written record should include (1) a summary of what was said, (2) the course of action the fire fighter needs to take to correct the problem, (3) a target date for completing the action, and (4) any decisions or agreements made between the fire officer and the fire fighter. The officer then signs the reprimand and asks the fire fighter to do the same. If the fire fighter refuses to sign, acknowledging that he or she has received the reprimand and understands its contents, the officer should sign and note that a copy was presented to the fire fighter who refused to sign. Remember that the signature of the fire fighter is not an admission of wrongdoing, it is simply an acknowledgment that a copy of the written reprimand was received and that its contents are understood. A copy of this reprimand is generally given to the employee, the union steward, and the personnel office. The fire officer should also keep a copy in his or her files. This step in the disciplinary process is often covered in union contracts. When this is true, the fire officer must be careful to follow the procedures specified in the contract.

Suspension is the act of relieving the fire fighter from all duties for a given period of time. During suspension, pay is reduced or eliminated. Benefits, however, generally continue. Suspension is used in serious offenses and reinforces the fact that fire fighters must behave appropriately as a condition of continued employment. The length of the suspension should correspond to the seriousness of the misconduct. The fire officer again must clearly communicate the reasons for the suspension to the fire fighter. In addition, the officer should offer suggestions for improving activities and should ask the fire fighter what he or she plans to do to correct the actions and attitude. The entire process of the suspension should be carefully documented in case a grievance or discrimination suit is filed. Suspension tends to be an effective disciplinary action in most cases. Repeated suspensions, however, lose their effectiveness and often signal a more serious problem than can be solved with this type of discipline.

Discharge is the highest form of disciplinary procedure and generally takes place after fire fighters refuse to change their conduct after repeated disciplinary actions, including suspensions. There are a few instances where the conduct of a fire fighter is sufficiently dangerous to warrant immediate dismissal or where the illegal nature of the activity is such that it warrants immediate dismissal if not prosecution. (For example, a fire fighter who commits arson.) The process of discharging a fire fighter is generally a tedious one. Discharge cannot take place until all facts have been gathered and examined. These facts must be considered calmly rather than emotionally. In addition, not only the immediate supervising officer, but other officers as well, need to be involved in the process. It is necessary to be as objective as possible in these instances.

Most labor contracts have a section dealing with discharge procedures. Fire officers must become familiar with these procedures and follow them correctly. Too often fire officers and departments are intimidated by the detail in union contracts and they avoid disciplinary actions. This must not happen, as it is a way of condoning and supporting misconduct. As long as the fire officers follow the procedures specified, appropriate disciplinary actions can be taken, even with a union contract. In fact, the presence of the contract makes discipline easier in some ways, because the exact steps are already outlined and agreed upon. Remember that failure to perform adequately in the fire service can result in injury or death to other fire fighters or civilians and cannot be tolerated.

Although it is a harsh action, it is imperative to discharge a fire fighter who refuses to follow the accepted performance standards. The fire department, other fire fighters, officers, and the community benefit from the action.

Discipline can be a motivating technique if it is performed in a constructive manner. Corrective actions and setting up a plan a fire fighter can use to improve performance assist in personal development. Motivation and discipline go hand in hand to improve individual performance and the overall atmosphere and operation of a fire department.

SUMMARY

Motivation and discipline are complementary management responsibilities that have a tremendous effect on the morale and performance of the members of the fire department. It is not always easy to motivate, but it is important to try. Motivation can be performed at all levels and results in a positive attitude. A department with high morale and pride in its professional performance is more productive and effective. It is also a pleasant place to work. The efforts required to motivate are worth the results.

Discipline can be used to guide individuals toward doing things the right way rather than as a negative tool of last resort. It demonstrates to every member of the department that correct completion of tasks is valued and no one is entitled to do less or to perform poor-quality service. Discipline ensures the maintenance of uniform minimum standards of service on the part of fire fighters and officers. Progressive procedures should be followed in disciplinary cases, matching the severity of punishment to the severity of infraction. Early mild disciplinary procedures can often prevent future serious actions that require severe, and difficult, disciplinary actions.

Motivation and discipline are ongoing activities that tend to occur in small events. Mentoring, on the other hand, is a long-term process that focuses on the overall career development of individuals. A proven approach, it is becoming more prevalent in the public sector. Together these three systems can improve the operation of a fire department and the fire service in general. Fire service administrators will utilize these approaches throughout their careers and need to develop competency and comfort in dealing with them.

CASE STUDIES

Case #1

A large city recently named a new fire chief to replace the retiring chief who had held the post for 16 years. This is the first chief to be selected from outside the department ranks. One of the challenges facing the new fire chief is low morale. A number of issues need to be addressed; however, many of these will take time and the fire chief wants to take some type of immediate action to demonstrate his intent to be responsive and make the department a better place to work.

During the first officers' meeting, a captain points out that a major source of irritation is the type of uniform shirt presently worn by fire department members. It is a formal, uniform dress shirt that is starched and pressed. Members are required to wear badges and name identification along with collar insignia. Many department members feel this type of uniform shirt is inappropriate for station duty or emergency response activity.

Recognizing the opportunity to make a gesture that could help improve morale, the chief orders open-collar sports shirts of soft material embroidered with insignia and rank. These are distributed to the fire fighters at no charge. The shirts become a standard part of the work uniform, and the fire fighters are pleased with the comfort of their new shirts.

The state labor bargaining representative, however, objects to this action because the new shirts are not part of the standard uniform that had been negotiated under the previous labor agreement. The uniform standard was part of the negotiation, and in order to make any changes the chief must confer with the appropriate union representative. It is determined that the chief has violated the terms of the contract.

1. How should the fire chief have approached this?
2. Did the experience improve the fire fighters' morale?
3. How will this experience affect the fire chief's motivation to improve things?
4. Could the fire fighters have done something to change what happened?
5. What could have been done to avoid the problem?
6. Is the real point of this incident the new shirts? What is the point?
7. What impact might this incident have on future actions of the new chief?
8. How would you, as a fire fighter, feel about what happened? How would you, as a fire officer, feel about what happened? How would you, as the fire chief, feel about what happened?

Case #2

Chief Jeffreys, part-time chief of the Volunteer Fire Department, was recently informed by the village treasurer that he will now be responsible for preparing and operating the fire department's annual budget. Significant cuts in the treasurer's office payroll budget have forced the layoff of one clerk, causing the treasurer to assume an additional workload. In past years, the chief has submitted a "wish list" to the treasurer, who then would prepare the budget in draft form and submit it to the chief for comments and approval. If the treasurer had deleted specific equipment from the request or reduced the amount of a previous allocation in certain line items, the chief and treasurer would meet to discuss it and attempt to reach a compromise. Throughout the next fiscal year, the treasurer would operate the budget and provide the chief with a tally of how much money remained in the fire department budget. Without assistance from the treasurer, the chief is faced with a monumental challenge since he is not familiar with municipal budget preparation or operation.

After considering several alternatives, the chief decides to call a friend who is an assistant chief with a large metropolitan fire department to ask for advice. The assistant chief expresses interest in helping, but admits that he does not have the training or experience to work on budget matters either. His department employs an administrative officer, Lieutenant Daru, who is responsible for numerous administrative functions of the department, including preparation and operation of the budget. The assistant chief arranges for Chief Jeffreys to meet with Daru the next day.

The first meeting is quite productive. The lieutenant provides samples of fire department budgets and explains in detail how to begin the task of organizing and preparing an annual budget. During the next month, the chief meets with Lieutenant Daru in person three times and calls several times on the telephone. Finally, Chief Jeffreys completes the working draft and visits with the lieutenant to request his review and comments. Daru compliments the chief on his budget preparation and encourages him to put it into final form for submission. He offers to teach the chief how to operate the annual budget once it is approved.

During their conversation the chief mentions that he is in the process of writing bid specifications for new department turnout gear and is not familiar with some of the safety requirements. Lieutenant Daru states that his department has just purchased new turnout equipment and that he will provide a copy of their specifications and help the chief prepare them for his department. Upon hearing this, Chief Jeffreys suggests that they explore the possibility of cooperating jointly on future equipment specification preparation and purchase since both departments could benefit from volume ordering.

1. Did the difference in rank affect this mentoring relationship? Why or why not?
2. What were Chief Jeffreys' goals upon entering the mentoring relationship?
3. How did the goals shift? Why?
4. Is the process becoming more formal?
5. How can Lieutenant Daru benefit from Jeffreys' suggestion to explore the possibility of cooperative purchasing?
6. Why was Lieutenant Daru willing to work with Chief Jeffreys?

Case #3

Captain Eddie was recently promoted to assistant chief in charge of fire operations. Eddie has been with the department for 9 years full-time and prior to that was an on-call fire fighter for 5 years. He has expressed confidence in his ability to handle the job and feels comfortable with his new position.

Following his first month in the new position, Chief Marks requests a meeting with him to discuss a recent property annexation to the community that will require the extension of fire and emergency medical services to the new area. The property includes approximately 95 acres of bog land, and during the meeting the chief requests that Eddie evaluate the potential impact a fire in the bog would have on the department's resources.

Knowing very little about bogs and nothing specific about fighting bog fires, the assistant chief begins the search to locate procedures for attacking such a fire or someone who has experience in this area. Within a few days, Eddie learns that Fire Chief Becker, who heads a small full-time department in a community 30 miles away, has considerable bog fire fighting experience gained through his association with the fire service in another state. Eddie contacts Chief Becker and requests his advice in addressing the bog problem.

Chief Becker invites Eddie to visit with him so that they can discuss the details and develop a plan of action. The chief generously shares his experience and knowledge and suggests several methods of fighting bog fires. Following the initial visit, Assistant Chief Eddie prepares a plan for attacking a bog fire that is compatible with the department's standard operating procedures.

Over the next 2 months, Eddie visits with Chief Becker three times and requests his input into the plan of action that is to be presented to Chief Marks. During the process, they develop an excellent working relationship and Eddie begins to trust Becker's judgment and advice.

Shortly after Assistant Chief Eddie receives approval of the plan from Chief Marks, the department is called to respond to a fire in the bog area. Once on the scene, Eddie discovers that the fire will be difficult to extinguish since it apparently has been burning undetected for some time. All of the department's resources are committed to the fire, and mutual aid companies have been summoned to assist in fighting the fire or for standby duty. Chief Becker's department is notified of the fire and he responds, along with a crew and one grass fire vehicle. Upon arrival, Becker locates Eddie and offers his assistance in organizing the fire attack. A grateful Eddie accepts the offer and asks Becker to evaluate the situation and suggest a strategy that would reorganize the fire fighting resources in order to initiate a vigorous attack. Becker suggests an alternative approach to attacking the fire, and Eddie takes immediate steps to initiate the changes that Becker suggests. Although Chief Becker does not assume a command role, he remains as a close advisor throughout the operation, offering guidance and encouragement the entire time.

Four days after the fire, Chief Becker telephones Eddie to discuss the bog fire and suggests several modifications to the combination fire department's operating procedures that would expedite the process of locating the seat of a bog fire and move fire fighting forces to the scene faster. He also praises Eddie for his command and coordination efforts during the incident and encourages him to continue to refine the bog fire plan of action.

One month later, the combination fire department responds to a second bog fire. Not wanting a repeat of the problems experienced during the previous bog fire, Eddie requests that Chief Becker be called to the scene. This time, however, Assistant Chief Eddie implements the revised plan of action that incorporates the suggestions Becker had made to him following the last bog fire. Once on the scene, Chief Becker observes the fire fighting operations but does not offer any advice to Eddie. The new plan is remarkably effective in controlling the spread of the fire until extinguishment is accomplished.

Chief Becker is impressed with the fire control activities of the department and informs Chief Marks of the outstanding job Eddie has done in preparing the bog fire plan of action. He encourages Eddie to pursue the development of other pre-fire planning projects and suggests that they meet on a regular basis to discuss Eddie's progress.

1. When did Eddie develop trust in Becker's experience and advice? How do you know?
2. The elements of this mentoring case study include motivation, trust, encouragement, and problem solving. Identify them.
3. What is the difference between Becker's role in the first and second bog fires?
4. Does it matter that the two officers are not in the same department?
5. Why do you think Eddie did not go to Chief Marks for assistance?
6. Can mentoring be applied to technical situations?
7. How does mentoring differ from calling in a consultant to assist with a technical problem?
8. What impact is this experience likely to have on Eddie as he faces future challenges?

Case #4

Willie Marshall is a paid on-call fire fighter/EMT who has been a member of the fire department for 7 years. Over the past 3 years, he is second only to the chief in the number of times he responds to emergency calls. Because his occupation is carpentry and wood finish work, he was able to make a significant contribution to the recent station remodeling project. In fact, he personally built all of the new cabinets in the radio room and the new kitchen. Although he is quite dependable and well respected by his officers and peers, he has a tendency to be quite independent and, at times, can become arrogant.

In January, an unusual snow and ice storm descends on the community, paralyzing the area for over 10 hours. Because the storm is unexpected, some residents are left stranded in their cars along many of the roadways. In addition, numerous motor vehicle accidents occur because of the road conditions. The fire department responds to assist with a minor accident that has occurred just a half block from the Number 1 station. Upon arrival, the captain in charge determines that no EMS or fire services are required; however, a hazardous condition exists because the roadway is coated with ice and partially blocked by the two vehicles involved in the accident. The local law enforcement agency informs the department dispatcher that there are no police officers available and that it will be approximately 2 hours before anyone can respond to take a report.

The captain decides to return the EMS unit to the station and keep the quick-response vehicle and its crew of three (including Willie) at the location to protect the intersection and accident scene, and provide a warning to other motorists. He orders the crew to set up road flares about 50 yards in all directions from the incident and requests that they assist other motorists by directing them around the site. Willie immediately objects to the order and informs the captain that he is "no cop" and therefore will not remain on the scene and perform law enforcement tasks. The captain insists that he stay, and Willie becomes quite loud and somewhat confrontational. He refuses to obey the order and walks back to the station.

After the accident scene is cleared, the quick-response vehicle returns to the station and the captain immediately meets with the assistant chief to discuss this obvious case of insubordination. After hearing the details of what had transpired at the incident, the assistant chief responds that although Willie's actions were wrong, he believes that no action should be taken. He tells the captain it has been his experience that in situations like this, "if you don't say anything or take any action, the condition will correct itself." Although Willie is in the station at the time, he ignores the captain and continues to discuss his actions on the call with other members of the department.

1. Will the situation "correct itself"?
2. What message does this incident give the other fire fighters?
3. Has the fact that Willie is a very active member made a difference in the discipline—or lack of discipline? Should it?
4. What action should the captain take? Why?
5. What effect should discipline have on Willie's future actions?
6. Is there any indication that there may have been minor infractions of procedure by Willie in the past? If so, how would you assume they were dealt with by the officers?

Case #5

The members of B shift have worked together for over 5 years and have become a cohesive fire and EMS response unit. Each knows his or her duties well, and in many emergency situations they require little or no supervision. In fact, a considerable rivalry has developed between B shift and the other two shifts. The assistant chief in charge of the shift has a close personal relationship with a number of the shift members.

Over the past 6 months, the overall discipline of the shift has deteriorated to the point that "freelancing" has become a standard method for addressing many situations. It began with trivial things at the station such as daily maintenance chores and has now progressed to conditions affecting emergency response. For example, Big Tex, a stout 240-pound fire fighter/paramedic, has always had trouble donning an SCBA while responding on Engine 4. He complains that the SCBA and mounting bracket are installed too close to the side of the personnel cab wall and therefore he is unable to perform this task. His captain evaluates the situation and informs Big Tex that no fire fighters on the other shifts are having this problem on Engine 4. The captain jokingly tells Tex to lose a few pounds and the problem will correct itself. Instead of following the captain's advice, Tex decides to modify his own in-station response actions by removing the SCBA from the holding bracket and placing it on the diamond-plate running board beside his turnout gear. The captain observes the change and informs Tex that it is not a safe practice to lay the SCBA in that position. Tex ignores the captain and continues to follow the same procedure for the next 2 months.

One evening, Tex leaves the station on a brief errand. During his absence, a fire alarm is received and Engine 4 is dispatched. In preparation for response, the captain forgets to remove Tex's SCBA and turnout gear from the running board, as he had done on two previous occasions. When Engine 4 turns at the first intersection, the SCBA and turnout gear are thrown to the pavement, causing irreparable damage to the SCBA. A passerby observes this incident, gathers the equipment, and returns it to the station. When informed of the problem, the captain is infuriated and immediately orders a 3-day suspension.

1. Is the 3-day suspension justified?

2. Can the suspension be enforced? Should it? Why?
3. What actions should the captain have taken previously?
4. How do you think the other members of the shift will react to the suspension?
5. What is an appropriate disciplinary action?
6. Who should be disciplined? Should it be more than one person?

Case #6

Sally Smothers is a part-paid fire fighter/paramedic who has been working for the department for 5 years. Smothers has a good performance record and is respected by both her officers and peers.

Last Wednesday, while at a car fire, Smothers fails to don her protective hood, as required by the department's standard operating procedures. The assistant chief, who arrives on the scene after the engine, observes her and physically pulls her away from the car, instructs her to stay with the engine, and says he will deal with her when they get back to the station.

At the station, the assistant chief confronts her on the engine room floor and calls her derogatory names. He also says she "is a dumb female" who doesn't have what it takes to be a fire fighter, proving that "women are not qualified to be fire fighters."

The following morning, fire fighter Smothers requests an immediate meeting with the chief to discuss the incident.

1. What disciplinary actions should the chief take against the assistant chief?
2. What was inappropriate about the assistant chief's actions?
3. How should the assistant chief have handled Smothers' failure to don her protective hood? What disciplinary action did fire fighter Smothers deserve for her actions on the fireground?
4. What disciplinary action should now be taken against fire fighter Smothers?

ENDNOTES

1. Buddy R. Silverman, "The Merit Pay System: Prognosis," in *Review of Public Personnel Administration,* 2 Summer 1982, pp. 29–34.
2. Carolyn Ban and Harry C. Redd, "The Status of the Merit System: Perceptions

of Abuse in the Federal Civil Service," in *Review of Public Personnel Administration,* 10 Summer 1990, pp. 55–72.

3. Brent Steel and Rebecca L. Warner, "Job Satisfaction among Early Labor Force Participants: Unexpected Outcomes in Public and Private Sector Comparisons," in *Review of Public Personnel Administration,* 10 Summer 1990, p. 4.
4. Gregory Lewis, "Pay and Job Satisfaction in the Federal Civil Service," in *Review of Public Personnel Administration,* 11 Summer 1991, pp. 30–31, passim, and Dennis L. Soden, "Esprit de Corps among Army Corps of Engineers Executives," in *Review of Public Personnel Administration,* 10 Summer 1991, pp. 50–54, passim.
5. Cole Blease Graham, Jr., and Steven W. Hays, *Managing the Public Organization,* 2nd ed. (Washington, DC: Congressional Quarterly Press, 1993).
6. Ibid., p. 172.
7. Ibid.
8. Ibid., p. 167.
9. Victor Vroom, *Work and Motivation* (New York: Wiley, 1964).
10. Sterling J. Livingston, "Pygmalion in Management," in *Harvard Business Review on Human Relations* (New York: Harper and Row, 1979).
11. Chunoh Park, Nicholas P. Lovrich, and Dennis L. Soden, "Testing Herzberg's Motivation Theory in a Comparative Study of U.S. and Korean Public Employees" in *Review of Public Personnel Administration,* 8 Summer 1988, pp. 40–60, passim.
12. David Clutterbuck, *Everyone Needs a Mentor: How to Foster Talent within the Organization* (London: Institute of Personnel Management, 1987), p. 7.
13. Linda Phillips Jones, *Mentors and Proteges* (New York: Arbor House, 1982), pp. 70–77, passim.
14. Graham and Hays, *Managing the Public Organization,* p, 148.
15. Ibid., p. 149.

REFERENCES

Ban, Carolyn, and Harry C. Redd. 1990. "The Status of the Merit System: Perceptions of Abuse in the Federal Civil Service." *Review of Public Personnel Administration.* 10, Summer, pp. 55–72.

Berkley, G., J. Rouse, and R. Begovich. 1991. *The Craft of Public Administration.* Dubuque, IA: Wm. C. Brown Publishers.

Blake, Robert, and Jane Mouton. 1964. *The Managerial Grid.* Houston: Gulf Publishers.

Clutterbuck, David. 1987. *Everyone Needs a Mentor: How to Foster Talent within the Organization.* London: Institute of Personnel Management.

Collins, Nancy W. 1983. *Professional Women and Their Mentors.* Englewood Cliffs, NJ: Prentice Hall, Inc.

Graham, Cole Blease, Jr., and Steven W. Hays. 1993. *Managing the Public Organization.* 2nd ed. Washington: Congressional Quarterly Press.

Hersey, Paul, and Kenneth Blanchard. 1982. *Management of Organizational Behavior.* 2nd ed. Englewood Cliffs, NJ: Prentice-Hall, Inc.

Herzberg, Frederick. 1968. "One More Time: How Do You Motivate Employees?" *Harvard Business Review.* 46:53–62 (January/February).

Jones, Linda Phillips. 1982. *Mentors and Proteges.* New York: Arbor House.

Kram, Kathy E. 1985. *Mentoring at Work: Developing Relationships in Organizational Life.* Glenview, IL: Scott, Foresman and Company.

Lewis, Gregory. 1991. "Pay and Job Satisfaction in the Federal Civil Service." *Review of Public Personnel Administration.* 11 Summer. pp. 17–31.

Lussier, Robert. 1989. *Supervision: A Skill-Building Approach.* Homewood, IL: Richard D. Irwin, Inc.

Murray, Margo, and Marna A. Owen. 1991. *Beyond the Myths and Magic of Mentoring.* San Francisco, CA: Jossey-Bass.

Novato Fire District. 1991. "Career Development Guide." Novato, CA: Novato Fire Protection District.

Olfshefsky, Dorothy. 1990. *Personal Interview.* American Society for Public Administration National Conference. Los Angeles, CA.

Park, Chunoh, Nicholas P. Lovrich, and Dennis L. Soden. 1988. "Testing Herzberg's Motivation Theory in a Comparative Study of U.S. and Korean Public Employees." *Review of Public Personnel Administration.* 8 Summer. pp. 40–60.

Rue, L. W., and L. L. Byars. 1990. *Supervision: Key Link to Productivity.* Homewood, IL: Richard D. Irwin, Inc.

Silverman, Buddy R. 1982. "The Merit Pay System: Prognosis." *Review of Public Personnel Administration.* 2 Summer. pp. 29–34.

Soden, Dennis L. 1990. "Esprit de Corps among Army Corps of Engineers Executives." *Review of Public Personnel Administration.* 10 Summer. pp. 41–54.

Steel, Brent, and Rebecca L. Warner. 1990. "Job Satisfaction among Early Labor Force Participants: Unexpected Outcomes in Public and Private Sector Comparisons." *Review of Public Personnel Administration.* 10 Summer. pp. 4–22.

Vroom, Victor. 1964. *Work and Motivation.* New York: Wiley.

Zet, Michael G. 1984. *The Mentor Connection.* Homewood, IL: Dow Jones-Irwin.

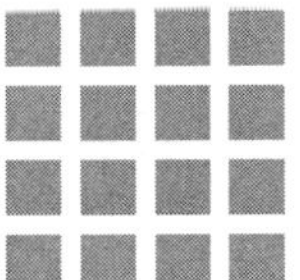

CHAPTER 7

Interacting with Other Public Administrators

J. D. Williams states: "Public administration is fundamentally the business of people—the people who work in the public sector and need to be turned on to their callings, and the people they serve who deserve responsive and responsible government."[1] This chapter considers these people and their interaction with the fire officer. No public service unit exists in isolation in today's complex world. Increasingly, interaction with fellow professionals is a key element of "getting the job done." The field of "globalism" and the concepts of "interconnectedness" are based on the realization that in order to address the challenges facing the public sector today, it is necessary to consider the greater (world) view of problems, issues, and potential solutions, and then to link with other appropriate entities in an attempt to provide comprehensive, manageable solutions.

Other chapters in this book have dealt with the more traditional aspects of working with subordinates, leading, mentoring, motivating, and disciplining. This chapter focuses on relating to other professional public administrators who serve in equal or superior positions to fire officers and chiefs in the organization.

The fire chief and other officers are public administrators in charge of providing a public service—fire protection—to the citizens of the community. It is important that fire officers be conscious of their status as professionals in a managerial field and interact with other professionals. These interactions take many forms, ranging from official scheduled meetings to informal discussions during daily activities. Each of these interactions is important; together they create the reputation of the fire department, department personnel, and the fire service.

Although the term "professional fire fighter" sometimes denotes a paid employee or a member of the union, a fire service manager or administrator is defined as a professional in a broader context. A professional has knowledge and skills in a given area that can be shared and evidenced in work-related activities, *regardless* of paid, on-call, or volunteer status. Fire chiefs of volunteer departments may interact more informally with other administrators, but they should maintain a professional attitude.

Fire fighters are "professionals" in the field of fire suppression and prevention; they have the necessary knowledge and skills to perform a specialized service function. Fire service managers have added the professional skills of administrators. A fire officer may have come from the ranks of the "professional fire fighter," but as an officer he or she is part of a larger profession of public administrators.

THE ROLE OF THE DEPARTMENT HEAD

The fire chief is the head of the fire department. However, public administrators, including those who are members of the fire service, often fail to consider the fire chief in the same light that they do other department heads. Certain physical indicators perpetuate this, including the uniform and the location of the chief's office in the fire station, a building separate from that of other department heads. Yet the fire chief has (or should have) the same responsibilities as other department heads.

The department head is the leader of a unit. For the fire chief this means more than being in command on the fireground. The fire chief is responsible for the daily operations of the department, is the leader of department personnel, and reports to superiors on behalf of the department.

In the past, the best fire fighter often became the fire chief. Unfortunately, this also meant that promotions were given to those who were not always qualified to fulfill their new duties as administrators. Fire scene command becomes less important as an individual advances up the officer ranks in a larger department. In a smaller department, where there are only two or three levels of officer rank, fireground command remains an important part of the job for a longer period of time. In these situations, however, the responsibilities of upper-level officers are often even more demanding, because these officers simply take on the administrative duties of their new positions in addition to their fireground command responsibilities.

A progressive fire chief grooms subordinate officers to take over key aspccts of department operations. Given the degree to which the chief must work with entities outside the department, duties associated with standard response to fire and EMS calls often need to be shifted to junior officers. This not only frees the chief to fulfill other demands, but also prepares these officers to take on increasing responsibilities, such as planning and budgeting. The delegation of fireground command responsibility to officers further down the chain of command provides an opportunity for those individuals to develop professionally.

When there are sufficient personnel to manage an average call, an effective fire chief is seldom on the fireground. If personnel arc properly trained and promoted according to their potential, the chief will be able to leave much of the standard emergency response activity to subordinate officers. The primary focus of leadership for fire chiefs is the daily administrative operation of the department. This may not be possible in all-volunteer departments, but members of volunteer departments need to recognize the additional administrative duties of the fire chief and allow time for their completion by sharing the burden of other activities.

Although the fire chief can delegate some functions of the office to subordinates, the chief is held accountable for the execution of these duties. He or she is responsible to administrative superiors such as the city manager or safety director, a political governing body such as the city council or board of commissioners, and the citizens for carrying out the tasks of the department. These include administrative functions as well as emergency response activities. A fire chief who overspends on equipment because of improper bidding procedures has failed to manage the department as has one who fails to train fire fighters for effective emergency response.

The fire chief has administrative duties such as preparing and managing the department budget, preparing department plans, equipment acquisition, personnel management, evaluation, and report writing. The chief's activities focus more on administration than on emergency response. Even in volunteer fire departments, the fire chief is usually responsible for all of the nonemergency response functions like record keeping, budgeting, and planning for the department, as well as recruiting, training, and motivating personnel. The future development of the fire service will require fire officers to think of themselves as managers and administrators and to acquire the skills to perform the broader organizational tasks of public administrators.

The fire chief is responsible for all department operations, not just what happens on the fireground. The chief must evaluate how the inspection program is proceeding and whether the information gathered is being used to preplan the structures and train the fire fighters properly. The chief must maintain a macro-level or comprehensive view of the fire department, understanding how various activities are integrated with one another.

Facilitating cooperation and integrating activities to meet the department goals are the most important elements of fire officers' responsibilities. To accomplish this, fire officers utilize various management and leadership skills on a daily basis. As individuals are promoted to officers, the demand for managerial and administrative skills increases. Departments must identify individuals who will make good managers and make managerial potential a basic requirement for promotion to the officer ranks.

INTERACTING WITH PROFESSIONAL COLLEAGUES

As an officer advances through the ranks, he or she must acquire the knowledge and skills necessary to carry out the administrative responsibilities of the new positions. Just as a fire fighter must learn pump operations and hydraulics, an officer must learn planning and resource allocation—through mentoring, formal education, and continuing education. Even if one has been a successful administrator for a number of years, continued professional growth and development enable one to do the job better. This is one of the functions of professional organizations and associations.

Professional associations have much to offer fire service officers. Participation, when done professionally and with an understanding of the purpose and operation of the association, can enhance a current or future administrator's professional development. Contributing to a professional organization is one way of contributing to one's own career. Taking advantage of what an association has to offer takes work, however. Officers who gain most from participation in a professional association are also those who give most to the organization.

Professional Organizations

Numerous organizations have formed around professions or types of professional activities. Involvement in professional organizations can be a rewarding experience. The fire officer should be involved in a number of these, for very different reasons. When deciding which organizations to join, consider what they have to offer and how you can contribute. These organizations may serve a region, state, or national constituency. Each has its own mission, and it is necessary to understand an organization's mission to determine the degree to which it serves your needs as a professional.

The fire service is a tradition-bound organization with a number of fire-service-based organizations offering various services to members. As administrators and managers, however, fire officers should recognize that fire-service-based professional organizations are not the only ones that can assist their professional development.

For example, The American Society for Public Administration (ASPA) is a major international professional organization aimed at developing the skills of public administrators and advancing the field of public administration. Although this is not a fire-service-based organization, fire officers are considered public administrators and are eligible for membership. This organization offers workshops at its national conferences that provide the opportunity to share innovations and research results. One of the newer sections of ASPA membership is the Section on Emergency Management, which includes individuals involved with hazardous materials response, disaster response and mitigation, emergency response planning, and exercise, as well as fire service administration and response.

At the local level ASPA chapters meet on a monthly basis and have various programs of interest to members. These meetings offer opportunities to meet public administrators from different departments and governmental entities. Because public administrators share the same challenges and responsibilities, these meetings provide a mechanism for a water district manager to interact with a librarian and a police captain on shared issues like cost management. This type of interchange broadens one's view of the community and increases awareness of what is taking place in the public sector.

The International City Managers Association (ICMA) also offers opportunities to develop administrative and management skills. ICMA hosts workshops and produces a series of public management books to assist public administrators in the performance of their duties. Among their publications are *Emergency Management* and *Managing the Fire Service.*[2] This organization also maintains a large database of information that can be used when conducting surveys to compare services offered and costs of services rendered by small communities as well as large cities. A particularly useful database contains information about local fire departments, including a contact person and information on staffing, apparatus, and salaries.

Technological changes have motivated the growth of numerous public-sector professional associations.[3] One of the oldest is the International Association of Fire Chiefs, which was founded in 1893. (The International Association of Chiefs of Police was founded the same year.) This international organization serves its constituents through a variety of activities, including a major emphasis on residential sprinkler installation, education and training opportunities, and the opportunity to network and learn from departments in other countries.

Since 1896, the National Fire Protection Association has been a leader in the effort to protect lives and property from the hazards of fire. NFPA publishes over 280 nationally recognized codes and standards, including the *Life Safety Code*® and the *National Electrical Code*®,

FIGURE 7-1 Professional associations offer opportunities for networking and career development.

as well as producing numerous fire service training programs and public fire safety education materials. Over 63,000 fire protection professionals comprise the international membership of NFPA. Membership benefits include technical assistance from staff, opportunities to assist in the code-making process, and a free subscription to the organization's bimonthly magazine.

The purpose, organization, format, and consequent benefits vary with each professional association. While each has something to offer, it is often best for fire officers to be active in different organizations at various stages in their career. This enables individuals to stay current in all areas, but also concentrate on improvement and participation in those areas that are most appropriate to achieve one's career goals.

Annual conferences facilitate interaction with professional colleagues and provide a forum for sharing innovative ideas and solutions to common problems. They provide a neutral forum where alternative ideas can be presented and discussed. Although debates can sometimes become rather heated, it is important to remember that in a professional debate the arguments attack and defend *ideas, not individuals.* The ability to discuss and argue ideas without attacking personalities is part of being a professional. It is not necessary to agree with individuals in order to respect them professionally or to work with them.

Local Professional Associations

State, local, or regional associations offer some of the same opportunities as the national organizations. These local associations serve as a training ground for participation in the national associations and can provide a continuous update on local changes. Associations focused on special areas of interest include those formed to discuss fire prevention issues, department safety procedures, and fire investigation. The debates in these smaller meetings often mirror those at the national level, and individuals can participate and prepare themselves to take part in similar interaction at the national level.

These associations also serve as a networking framework. Other members can be called upon for information concerning a special challenge. Interacting at the meetings acquaints members with each other's styles and areas of specialization, and will help direct inquiries to the appropriate person. Asking for assistance and sharing expertise are professional development activities that benefit all parties involved. No one can be an expert at everything. Developing a network of resources to call upon for assistance enhances a manager's ability to use available expertise and avoid learning everything from the ground up.

This interaction is a two-way street. Members of networks are expected to share as well as request assistance and information. It is important to emphasize the professional nature of the relationship. A member of a professional network must be willing to participate and share knowledge and information in a professional manner, and to assist others when requested to do so.

One other operational aspect of networks needs to be clarified. The give-and-take of information exchange may not be equal between individuals, or **dyadic. Multiple-member assistance patterns** are stronger,

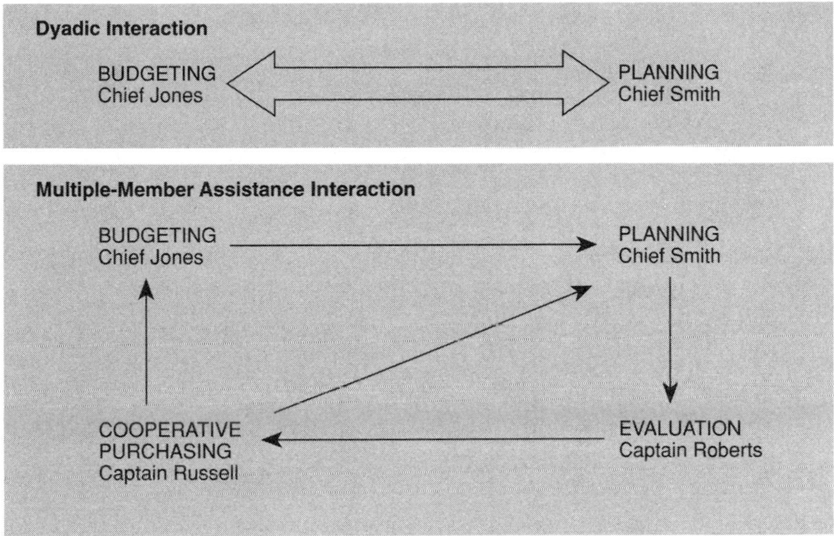

FIGURE 7-2 Dyadic interaction versus multiple-member assistance interaction.

more frequent, and increase the synergism of professional networks; they involve a group of individuals who help one another in various situations without requiring a direct exchange of favors. See Figure 7-2.

Note that with the multiple-pattern interaction one person or area of expertise may be called upon more than another. For example, Captain Russell may be asked about cooperative purchasing by a number of departments until a system is established in the region. In the short term, there may be an unequal burden of information sharing. In the long run, however, the exchanges even out. Moreover, as a fire officer shares information in an area of expertise, that individual also gains recognition from professional colleagues.

Local fire associations can also provide unified strength when facing a challenge affecting many or all members and their departments. For example, an association may take a position concerning an amendment to the state building code. Although individual departments and fire officers may also take a position on the issue, the association statement will generally have greater impact.

WORKING WITH OTHER DEPARTMENT HEADS

There is a general perception, even among public administrators, that the fire chief is not a real "department" head. This is based on a narrow perception of the department's activities. In addition to putting out fires, the fire department must perform the same administrative and management functions as other public departments: planning, budgeting, human resources management, communications, etc. Thus, the fire chief is also a department head[4] and must become involved in cooperative interaction with other department heads.

Increasingly, department heads are involved in trying to identify proactive ways to serve the community. Department heads are forced to be more innovative and to examine alternatives for service delivery, including cooperative arrangements with other departments such as equipment sharing, personnel cross-training, and integrating planning activities. These arrangements not only increase the capacity of each department, but they also integrate the fire department more fully into the center of government operations and provide department members with a more comprehensive understanding of what government service to the community means.

Facilitating this type of cooperative interaction requires effective communication. Department heads must be able to explain their operations, procedures, and needs in a manner that the other department heads can understand, and they must understand the concerns and goals of other department heads.

Fire department heads do not necessarily need to be familiar with every aspect of daily operations in the department; however, they do need to understand the purpose and goal of each unit. Obviously, the need for detailed understanding depends on the size of the department. In volunteer or combination departments, or even smaller paid departments, it is more important for the chief to be familiar with details of operation. However, as the departments grow in size and the various units become more distinct and function more independently, it becomes impossible for the chief to concentrate on the operational details of each subunit. The head of the fire department must be able to communicate the functions of each unit at an appropriate level of detail to manage effective cooperative interaction with persons outside the fire department.

For example, the planning board and a developer may be discussing line sizes and completion dates for hydrant installation in a new development. If the fire department is responsible for hydrant testing, a knowledgeable fire officer should be involved so that the testing can be scheduled and the department's input on preplanning is considered. Factors such as scheduling, water pressure and gallonage requirements, and placement of hydrants to facilitate access to the buildings will have a direct impact on fire department operations in an emergency and on daily operations. By discussing concerns relevant to department goals and operations, the fire officer can effectively communicate information that will enhance the completion of the new development.

The fire officer must have a thorough knowledge of the fire department and understand the purposes, operations, and structures of other municipal departments. This does not mean that a department head must understand everything about the operations of other departments, but he or she needs to understand enough to relate to their concerns and goals. *Thus, although some information concerning department responsibilities may not relate directly to the fire service, the more aware that fire officers are of the competing demands for service resources, the better able they are to explore alternative means of providing service to the community.* As Ronny Coleman, fire chief of Fullerton, California, has pointed out, "Many communities planning to adopt residential sprinkler ordinances have not succeeded because they failed to establish good working relationships with local water departments."[5]

Primary Forms of Local Government

Familiarity with the organization of local government is important for fire service officers. It is impossible to relate and interact with other professionals outside the fire department without understanding their position in the organization. In addition, when interacting with members of the fire service from other communities, a basic understanding of how their community is governed provides a much clearer understanding of the challenges they are facing. Understanding the similarities and differences will enable fire officers to more readily grasp ideas that could be used in their own departments and to understand the type of adaptations or limitations necessary for others to succeed.

Although local government comes in a variety of sizes, most follow traditional organizational formats. NFPA 1021, *Standard for Fire Officer Professional Qualifications,* identifies knowledge of various forms of local government important for fire officers. Accordingly, the following discussion of the primary forms of local government will assist in clarifying questions of organization.

NFPA 1021, Standard for Fire Officer Professional Qualifications Requirements relating to government structure, Fire Officer II:

3-7.1 *Describe the functions of other bureaus, agencies, and divisions of government and the need for interagency and intergovernmental cooperation.*
3-7.2 *Describe the law-making process at the federal, state/provincial, and local levels.*
3-7.3 *Identify the basic characteristics of local forms of government in the authority having jurisdiction.*

Fire Officer IV:

5-7 *Interpret authority of the branches of local, state/provincial, and federal governments serving the authority having jurisdiction.*

City Government

At the core of local governments are some type of elected public officials. These public officials are elected by the residents of the community to represent their interests in the governance of the community.

City government is primarily one of three types: council-mayor, council-manager, or commission. City councils are comprised of elected individuals who either represent specific wards or districts in the community or who are elected at large. At-large representatives are elected by all eligible voters in the community. Single-member districts or wards refer to subdivisions of the city, which have one representative elected from each. Most cities have single-member districts, some smaller cities still have at-large representation, and a few, like Dallas, Texas, have both. Initially, all councils were comprised of representatives elected at large. As communities grew and diversified, however, it was felt that minority groups were denied representation. In addition, certain parts of the city did not receive their share of city services. Thus, cities were divided, on the basis of population, into districts, precincts, or wards.

In a **council-mayor** form of government, the council members and the mayor are elected by the residents. The mayor serves as the head of city government and may or may not be head of council. In some cities, the head or president of council is elected by the members of council. This form of government is further categorized as either a "strong mayor" and "weak mayor" type.. In a weak mayor city, the mayor is primarily a figurehead and ceremonial leader. The mayor has little or no administrative power and does not take a very active role in running the local government. In a strong mayor setting, the mayor is a very active leader in city government and, in fact, runs the city as its chief administrator. In this instance, the mayor has considerable power over daily operations and frequently has a lot of influence with council as well. Often the strong mayor is also the head of council.

In a council-mayor form of government, the fire department usually reports directly to the mayor. The fire chief is the primary point of contact, and the chain of command becomes: council–mayor–fire chief–fire officers, etc. The fire chief may be called upon to report to council, but most of the interaction will be with the mayor. Thus, it is important to develop a good working relationship with the mayor. Since mayors often have little understanding of fire department duties and operations other than emergency response, it is important to communicate information about the department effectively.

Because mayors are elected, they are conscious of the demands and expectations of the residents. In addition, they are supportive of popular and effective services to the community. Thus, fire departments that report to a mayor should publicize public education and awareness programs and invite the mayor to community functions hosted by the department. This can enhance understanding and smooth working relationships.

The **council-manager** form of government is based on the principle that politics and administration of city government should stay separate. In this form of government, the members of the city council are elected by the voters, as in the council-mayor form. The mayor is also elected, generally at large, and serves primarily as a ceremonial leader and head of council. The mayor has no administrative duties in the city. Instead, the council hires a professional public administrator with the appropriate background experience and education to serve as the manager of the city. The city manager is a professional administrator, not a politician, and does not run for office. The duty of the city manager is to carry out the policies of council in the most efficient and effective manner possible. The council-manager form of government tends to be most prevalent in the southwestern and western parts of the country. Generally, areas that value professionalism in government have a higher number of council-manager governments.[6]

The fire chief generally reports directly to the city manager unless there is a safety director to whom police, fire, public health, and sometimes building inspection departments report (as well as EMS when that is an independent service). In this instance, the city manager is less concerned with public relations and more concerned with efficiency in operations and effective performance of duty. While the department would still benefit from inviting the mayor to community awareness programs, relationships with the city manager are enhanced more by professional presentation of information, accurate budgets, and well-thought-out, comprehensive plans. In addition, cooperative relationships with other department heads and awareness of local government issues tend to enhance the fire chief's standing with the city manager.

The fire chief may have to establish committees within the department to provide the necessary expertise for the various tasks assigned by the city manager. The chief may also choose to assign fire officers to participate in committees outside the department to coordinate tasks, represent fire department interests, and facilitate cooperation among municipal departments. Officers may also assist the fire chief with report writing or developing presentations for the city manager. Committees provide a formal system for gathering information from fire officers, provide a broader scope for officer activities, develop municipal relationships, and allow the fire chief to coordinate information gathering and engage in larger management and administrative issues.

A **governing board of commissioners** is not as prevalent as it once was. Cities that do have commission governments are usually smaller in size. Commission governments almost always elect members in an at-large election. Each commissioner is then responsible for the daily operation of one or more public departments. For example, on a three-member commission, the first commissioner might be responsible for public safety, including police, fire, and public health, the second might be responsible for roads, transportation, and water and sewer, and the third might be responsible for tax collection and social services. This system is a direct mix of political and administrative duties and thus is the opposite of the council-manager form. As cities grow in size, the size of government expands and the use of the commission form of government usually gives way.

In the commission form of government, the fire chief will interact most directly with the individual having jurisdiction over the fire department. This commissioner is responsible for the daily operations and usually has the responsibility of persuading the other commissioners to allocate necessary increases in resources. The fire chief should maintain open and honest communication with the commissioner in charge of the department. When requesting additional resources, the chief must provide the required information in a form that will help the public safety commissioner convince the other commissioners. It is important to build a positive relationship with each commissioner and to support their interests as well. Again, invited participation in public awareness events will help build a supportive attitude toward the fire department.

County Government

County government generally takes one of two forms, either commission or commission-manager. Counties were originally created to decentralize the state government. Where it was too difficult for people to travel to the state capital, counties were created to provide an outreach for state government so that people could do business with the state in a more effective and efficient manner. In addition, the state was able to run its programs more effectively by providing services through a locally based center of outreach. Thus, much of a county's responsibility is centered around the implementation of state programs.

County commissioners, like those of cities, take an active role in the administration of local government services. This system may be organized along service areas as described for cities, or they may be a combination of shared and individual assignment of responsibilities. For example, in some states each commissioner is responsible for the roads in his or her own district. Thus, the responsibility for statewide road improvement is shared among commissioners. However, each commissioner is then given primary responsibility for other department operations, creating a mixed set of responsibilities. In all cases, however, the final authority lies with the majority of the commission as a whole. The majority vote will dictate the outcome of a dispute.

Some counties are organized as a commission-manager form of government. These arrangements sometimes follow the city model of having the commissioners hire the county administrator or manager. In other counties, however, the county administrator is elected to the position by the public. This obviously negates the intent of the council-manager form of government to separate politics from administration. Instead, these counties simply believe that there needs to be one central locus of administrative control and that the citizens should identify the individual who is to hold that position.

Township Trustees

Many townships are governed by boards of trustees who are elected either at large or from districts. Most are elected at large. It must be remembered that even though an election might be held at large, meaning that all of the residents can vote for any candidate, there may be different "places" or "seats" identified, for which candidates must file. Thus, someone who wants to run for a trustee position might run for the third place or seat, competing for others who have filed for that race, and not be in competition with those who filed to run for the first or second seat.

Township boards of trustees function very similarly to commissions in that individual trustees have primary responsibility for various parts of government service. The operation or management of a township is often informal, as the size makes frequent communication easy.

In townships the fire chief should try to maintain positive working relationships with each trustee. Although daily operations and reports of activities usually go to one of the trustees, communication concerning plans, special activities, or resource uses should be provided to each trustee on a periodic basis. In order to ensure that the trustees understand fire department operations, it is a good idea to explain the different programs and to offer an open-house tour for all new trustees. In addition, it is helpful to include the trustees in an annual event. This helps build strong and favorable relationships between the department and the board of trustees.

Special Independent Districts

Special independent districts are the fastest-growing form of local government. Special districts are organized to deal with one specific problem or item of public service to consolidate resources among communities. Fire districts are special districts, as are school districts, hospital districts, and water and sewer districts. District boundaries are not always conterminous with local government boundaries. In other words, special districts often include areas that are part of a variety of different local governmental jurisdictions. Special districts are often governed by an elected or appointed board. Where the board is appointed, it is generally the case that the local governmental jurisdictions served by the district select representatives to serve.

Special districts have all of the rights, privileges, and responsibilities of local governments. This includes, most importantly, the power to raise revenue through taxation and borrowing. Fire districts can raise revenue, contract for services, apply for assistance, set policy, and enter into binding contracts with other local governments. Fire districts are found in areas where local government is not large, wealthy, or strong enough to provide fire protection services. This is especially true in rural areas where fire districts often cover more than one local jurisdictional area.

The feasibility of expanding the fire district concept to enable cooperative consolidation of service areas and maximize resources is gaining considerable attention. In times of financial scarcity, sharing resources is an attractive concept. Fire districts have and continue to make a major contribution to public protection in the U.S. More interest, investigation, and discussion of this possibility is certain to take place wherever public revenues are decreasing.

COMMUNICATION: THE KEY TO EFFECTIVE INTERACTION

Communication is an essential component of every organization and every relationship. The use or misuse of communication can make or break a professional relationship. Positive communication enhances the stature of both the organization and the individual.

Positive communication means sharing information in the right way at the right time. Not all information will be good news. However, there are appropriate ways to share bad news that make the fire officer's role as the bearer of bad tidings more acceptable. It is always best to be open and honest about unpopular information and attempt to work out solutions with others. Avoiding a problem or withholding information until it is discovered accidentally is demoralizing for staff and may be interpreted as a misuse of power. Thus, praise, recognition, and compliments, coupled with judicious discipline and honest sharing of bad news, comprise "positive communication." Professional interaction with colleagues entails appropriate amounts, types, and tones of information sharing—of communication.

Managers and administrators are well aware of superior-subordinate formal communication, which follows an organization's chain of command. Generally, the aspect of informal communication within an organization is also well known to fire service officers. Formal and informal communication with other public administrators is not as well defined, however, due to the relationship in an organizational chart or the lack of a shared organization. Nonetheless, both formal and informal communication are warranted as the fire chief interacts with other fire chiefs outside the area and with department heads in the city, township, or county.

At the higher levels of an organization (department head and above) there is usually a higher degree of informal communication and information exchange that supports smoother operations in all units. Thus, after receiving formal communiqués, one department head may simply pick up the telephone and notify another department head of activities that will affect the second department. This enables the second department head to manage the impact of changes on his or her department. It requires clear understanding of confidential versus general information and the ability to interact formally and informally with professional colleagues who are potential information sources. Thus, the planning director might call the fire chief about an annexation proposal or about a building permit request for an industrial development.

Informal communication can be positive or negative. Negative informal communication is gossiping and unauthorized sharing of confidential information. It undermines fire department operations and authority as well as trust in the officers. Positive informal communication helps develop networks. Networking is an informal pattern of communication connecting individuals with others in the organization or with those outside the organization. In a fire department, an officer in charge of ensuring adequate water supply would network with the community water department administrators and officers in other departments working on similar projects. Providing the organization with a wide range of information allows a better exploration of alternatives. A well-networked organization is less dependent on small groups of upper-level administrators to make all decisions and to develop all alternatives. Networking provides a broad base of information and allows individuals to specialize and become experts on particular issues. This facilitates decision making because expertise is available within the department to provide the necessary foundation for making informed decisions.

Complete control over all aspects of communication in an organization is impossible. When deciding how and what to communicate, administrators should form broad outlines and guidance, leaving specifics to those most closely involved with specific technical aspects of information. Sometimes ambiguity in communication policy can be a positive strategy. By making informal decisions, administrators can get feedback and discussion to explore alternatives before finalizing choices. Adhering to a strict guideline weakens an organization's ability to adapt to new communication needs and respond to a need for informal information. In addition, strict communication guidelines discourage change and adaptation to changes in the community.

The physical location and layout of a department directly affect communication and information flow. The nature of fire department service often requires that parts of the organization be physically distant from one another. In addition, the fire department itself is often located away from other government departments. This means that the fire officers must be much more cognizant of communication patterns inside and outside of the department and work to ensure the smooth, timely flow of information.

In a volunteer or paid-on-call department, communication can be even more challenging, as it is not always possible to reach every member or every officer at the same time. In some departments, when something is changed, the changes are shared only with whoever is around at that point in time. This can result in confusion, misinformation, and a general feeling of frustration and disorganization. Sharing information is one way of letting the members know that their input is valued. It is crucial that communication be systematic in volunteer departments to ensure that everyone who needs the information gets it. The system of communicating information must be formal or consistent, even if the content of the information shared is not.

For example, if the state is going to offer free special rescue classes, the chief of a small volunteer department must let *all* officers or *all* department members know about the opportunity in a timely fashion. This can take the form of a phone call to *each* officer with instructions to "pass it on" to their assigned members or a page signaling "training opportunity—details with dispatch at the station." The important aspect of this communication is that it is not selective.

Other information may not need to be shared with everyone, but it is important to make sure that those who do need the information are given it. Communication challenges in smaller volunteer organizations are just as great as those in large organizations, sometimes more so, and just as critical.

INTERAGENCY AND INTERGOVERNMENTAL COOPERATION

Fire departments were originally established to deal with a specific local problem: fire. However, as communities grew and became more complex, the department activities expanded into public education, code enforcement, EMS, and, today, hazardous materials spill mitigation. These responsibilities are no longer isolated. Public education involves the participation of the public schools; code enforcement involves business and industry as inspections are performed; EMS works closely with the hospitals; and hazardous materials response is a complex, negotiated effort on the part of numerous agencies from three or more levels of government. Interagency and intergovernmental cooperation is critical to the function of the fire department in the 1990s and beyond.

Cooperation is defined as a common effort or association of per-

FIGURE 7-3 Interagency cooperation is essential to increasing community safety.

sons for common benefit. Intergovernmental cooperation refers to the collaboration of federal, state, and local governments. Interagency cooperation refers to the sharing of responsibilities among different agencies that have individual areas of responsibility. For example, the school system has responsibility for education, the police department has responsibility for protecting the community from crime, and the fire department has responsibility for protecting the community from fire. Together they can cooperate to increase community safety through school-based public education and training programs.

Given the scarcity of resources and the increasing complexity of our world, intergovernmental and interagency cooperation is the key to successful public management in the future. Federal programs are implemented at the local level and require state participation. Current issues have an extensive impact, requiring attention from a number of government agencies. SARA Title III, for example, requires considerable intergovernmental and interagency cooperation in order to meet the mandatory responsibilities.[7]

Title III of the Superfund Amendments and Reauthorization Act (SARA) of 1986 mandates local hazardous materials Emergency Response Planning and Community Right-to-know. (Public Law 99-499) SARA Title III requires the governor of each state to appoint a State Emergency Response Commission (SERC), which must then: (1) establish emergency planning districts, (2) appoint members to the Local Emergency Planning Committee (LEPC) in each district, (3) supervise and coordinate the activities of the LEPCs, (4) establish reporting procedures for industry, (5) establish procedures to process public requests for information, and (6) review local emergency plans.[8]

Manufacturing or storage facilities having on-site designated hazardous materials in excess of the established threshold limits must report their presence to the SERC, the LEPC, and the local fire department. The LEPC and the SERC are charged with ensuring that this information is then made available to the public under the right-to-know provision. The LEPC and the fire departments are charged with preparing an emergency response plan for the local emergency planning district, which the SERC must review and certify. Most of the implementation responsibility of SARA Title III rests with the LEPC in a complex intergovernmental relationship.

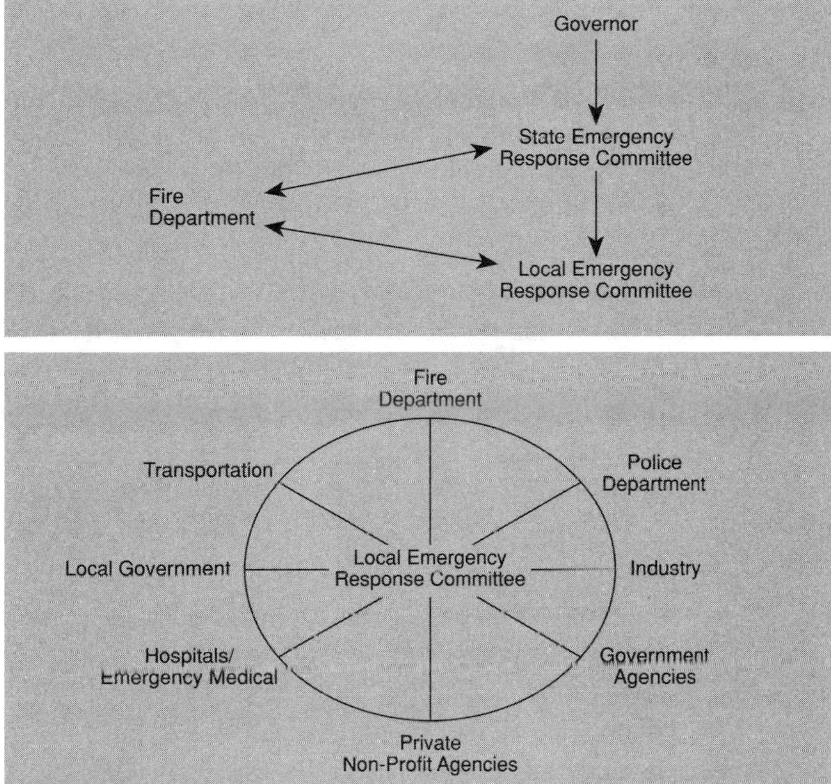

FIGURE 7-4 Implementation of SARA Title III requires considerable interagency cooperation.

SARA Title III is implemented at the local level so that information and plans are more readily available to and appropriate for first responders in a local hazardous materials situation. The SERC was asked to identify the areas designated as Local Emergency Planning Districts to ensure that their geographical boundaries matched those of other existing state programs. The federal law also dictates which groups will be appointed to the LEPC by the SERC. For example, each LEPC must have a representative from industry, law enforcement, health, etc. Thus, the nature of the cooperation between federal, state, and local government entities to fulfill SARA Title III requirements is obvious.

SARA Title III designates fire departments as primary first responders to hazardous materials incidents. Fire departments must use hazardous materials information from industries in their fire districts to preplan response procedures for a hazardous chemical release. Plans from the different fire departments in the district are integrated into the LEPC district-wide plan. Mutual response and mutual aid agreements are spelled out in this districtwide plan to assign responsibilities and actions of individual departments in advance of a hazardous materials incident.

Implementation of SARA Title III requires cooperation among many agencies and private industries. Too often emergency response planning focuses on the responsibilities of separate departments. This results in uncoordinated emergency responses that may fail to mitigate the incident as rapidly as possible. SARA Title III requires the coordination of all activities, regardless of agency affiliation, and mandates widespread participation in planning. The comprehensive LEPC's master plan incorporates each agency preplan and usually includes the entire agency plan as an annex. The cooperation prevalent under SARA Title III extends to private industry as well. Private business must be represented on the LEPC and SERC. Local facilities with emergency response plans must share that information with the local fire departments and the LEPC. This ensures cooperative rather than conflicting response activities. Under SARA Title III, fire departments must participate in two types of cooperative behavior, as must all other participants in the process: (1) agencies of like type, such as fire departments, must cooperate and (2) different agencies must cooperate with one another, e.g., the transportation service must cooperate with law enforcement in evacuation planning.

The training mandates of SARA Title III illustrate federal interagency cooperation between two separate agencies, the Occupational Safety and Health Administration (OSHA) and the Environmental Protection Agency (EPA). The *Federal Register* of March 26, 1989, identifies the OSHA-specific skill levels and corresponding training requirements for hazardous materials emergency responders. In the June 23, 1989, *Federal Register* EPA adopted those standards for all emergency responders identified under SARA Title III. The EPA also mandates that the training requirements designated by OSHA are required in non-OSHA states as well. This action created a universal training standard for emergency first responders in hazardous materials incidents at the federal level and illustrates how activities at the federal level can have direct impact on local fire departments.

SARA Title III is a good example of the increasing need for intergovernmental federal, state, and local coordination, as well as interagency and intergovernmental cooperation among fire departments, police departments, cities, counties, hospital districts, other local governmental entities, and even private industry.

This discussion has focused on one illustration of cooperation. Fire departments are going to be involved in these types of activities more and more frequently in the future. Consequently, fire officers need to be cognizant of changes in their communities and of activities taking place at the state, national, and even international level that may have a direct impact on their future operations.

ENVIRONMENTAL SCANNING

In the 1980s, a management technique called environmental scanning was developed to help community and organization leaders maintain an awareness of changes that might affect their operations. Environmental scanning is the technique by which fire officers identify and examine technical, social, and political information, inside and outside the organization, to determine current and potential trends that may affect the organization. This technique can be extremely valuable in the planning process, especially strategic planning, as the information can be used to prepare alternative strategies for dealing with the impact of such changes. The primary value of environmental scanning is preparedness.

Environmental scanning might be used to mitigate the local impact of an international activity, such as Operation Desert Storm. A progressive fire officer would relate concerns about federal plans to use National Guard reserve units in the Middle East and its potential impact on department staffing. By determining the number of department members involved in local National Guard units, the impact of activating the National Guard would be known and plans could be made to provide alternative staffing if necessary.

David Aaker presented a six-step approach to environmental scanning that he named the Strategic Information Scanning System (SISS).[9] The first step is to identify the information needs of the organization. Thus, participants in the scanning process should be familiar with the operations, concerns, and anticipated future decisions faced by the fire department. In addition, they need to understand how various actions in the community influence fire service operations. The importance of different types of activities is determined by answering three questions:

1. What will the probable impact be on the department?
2. What is the probability it will take place?
3. How much time does the fire department and its members have to respond if it does occur? (How long before the department is affected?)

The second step is to identify all potential sources of information. Professional networks and colleagues both in and out of the fire service can be extremely valuable in this process. In addition, professional and contemporary journals offer insight into current trends. Obviously, local and state publications detail decisions and changes in the economic and political environment of the department. These sources must be carefully monitored to ensure that the fire department is not surprised by new responsibilities or shifts in legislative requirements. A formal list of data sources makes officers aware of sources of information that should be checked on a regular basis. This list can be in two parts, the first identifying printed documents and the second, organizations and individuals. Then, if a specific change is noticed, such as rumors of a change in the state fire or building codes, officers know which publications to read and who to contact locally and/or at the state level in order to find out what is going on.

The third step is to identify all department or organization members who will be involved in environmental scanning. It is important to look beyond rank and identify anyone in the department who may have access to useful information sources. All fire fighters should participate as well as officers involved in department-level decision making and planning activities. The key is to involve two sets of individuals: those aware of environmental changes and trends and those involved in planning and decision making. Hopefully, there will be a great deal of overlap in these two groups. The optimal situation is one where fire officers involved in planning and decision making are also keenly aware of the nature of changes in the internal and external environment of the department. The process of identifying specific individuals and identifying expectations for their participation is crucial to the success of environmental scanning and applies to departments of all sizes. This step reinforces the importance of the process and identifies specific expectations of individuals. It makes individuals part of the team.

The fourth step is almost a continuation of step three. It involves matching the individuals identified in step three with the sources and areas of information identified in steps two and one. This lets everyone involved know exactly what they are to look for and what areas of information they are to scan continuously for relevant data. These assignments can be based on department responsibilities or personal interest. For example, the officer in charge of the hazardous materials response team and a fire fighter, not on the hazmat team, with a degree in chemistry and an interest in hazardous substances may both be part of environmental scanning and assigned to work specifically on areas that might influence department hazmat team operations.

The fifth step is identifying how and where the information is to be stored. Whether the people involved take notes, clip articles, write summaries, or even just meet once a week to report on their area and have minutes taken at the meeting, some systematic method of recording the data and keeping it in one place needs to be adopted. This ensures that the information can be referenced when making decisions. Periodic sharing can also help to identify trends.

The sixth and final step is also the most difficult. This involves using knowledge gained to make predictions about the future. Experts agree that the forecasting or predictive step is the most difficult. It relies on human judgment and reinforces the fact that "technology cannot substitute in this complex process for the trained mind and alert eye."[10] The predictions, however, generate ideas about the possible and probable alternative futures faced by the department. Becoming aware of these different scenarios enables the department to better prepare and can, in some instances, assist the department in shaping its future by being prepared for changes.[11]

The following three activities are the essence of the environmental scanning process and may be more appropriate for smaller departments. Olsen and Eadie reduced the process to three steps: (1) identify relevant factors, (2) forecast how the factors will change, and (3) assess the impact of the change on the department.[12] Identifying the people who will take part in this activity is, however, extremely important, as it elevates the process from an off-the-cuff activity to a formal part of the strategic planning process. (See Chapter 10, Planning.)

In a smaller community with a more stable environment, there are considerably fewer variables to be considered in the scanning process. In these instances, environmental scanning activities can be simplified and made more informal, reducing the time and effort involved. What makes environmental scanning a useful management technique is that changes that may affect the department are formally recognized and taken into account as the department prepares for the future. This formal recognition is the key to being prepared for changes. Departments of all sizes can benefit from being prepared to deal with changes in their communities. Increasing the level of awareness makes a tremendous difference in a department's ability to react positively to new challenges. Environmental scanning is a technique that facilitates preparedness.

SUMMARY

Fire service administrators must be able to interact positively with other public administrators. This entails developing communication and interaction skills necessary to disseminate and share information, the capacity to appreciate the concerns of non-fire-service public administrators, and an ongoing awareness of the community and environment in which the fire service operates.

This chapter has examined the expanding role of fire officers and offered some insights into how to prepare to be more effective in the future. In a service based on technical skills, interpersonal capabilities and general knowledge required in interactive settings are often foreign and uncomfortable. However, interactive communication and management skills can be refined and developed within the department.

The need for a broad-based knowledge and understanding of the community in which the fire department operates is increasingly critical for effective management of the service. Isolationism is no longer a viable political position in international affairs or economic planning; nor is it a workable approach for fire service officers. Fire departments whose leaders ignore the changes in the community and refuse to adjust methods of operation to meet the service demands and resource capacities of the community will have change forced on them. Fire departments have the capability of directing their future as long as the abilities of their officers are developed and utilized.

CASE STUDIES

Case #1

The City of Burton is a moderate-sized, blue-collar community located along the Atlantic seaboard. Since the mid-1970s, numerous manufacturing facilities have moved out or closed, causing many citizens to be unemployed. Several state and federal job retraining programs have been instituted to help the citizens recover and prepare for the future. As part of the revitalization effort, a number of commercial properties have been identified as prime locations for new businesses seeking to relocate to the city. Although these locations are in close proximity to major interstate roadways, there is no direct convenient access to them since most of the streets in the area are narrow, traffic-congested, two-lane corridors.

To remedy the situation and make these commercial properties more attractive, the city council authorizes the planning department to proceed with a study of the present land use in areas contiguous to the commercial properties in question, and to develop a comprehensive land-use plan that will contain solutions to the transportation access problem. After nearly 2 years of deliberation and study, the master land-use plan is unveiled before the council. The focal point of the report is a proposal for several of the two-lane streets to be widened to four lanes to accommodate swift access to the interstate road network. Much to the surprise of the fire chief, one of the streets to be widened is Fifth Avenue, on which Fire Station 3 is located. The revamping and widening of the street would involve taking 18 feet from the station ingress/egress driveway. The renovation would present major problems for emergency response vehicles as they respond or return to the station. In addition, the proposal does not provide for a traffic light at the intersection just east of the station. Since the station does not have drive-through bays, this means that fire fighters will be required to physically stop all traffic across three lanes in order to back fire apparatus into the station. In addition, the absence of a traffic light will prevent the use of a light block to stop traffic when leaving the station.

The fire chief is distressed by the proposed widening plan on Fifth Avenue and approaches the planning director with her concerns. The planning director is somewhat upset and questions why the fire chief didn't identify this problem during the study and before the public meeting. The fire chief explains that she was not aware that the study was being conducted since she has never been consulted about its impact on the fire department.

1. Who is at fault?
2. What could the fire chief have done to become part of the process?
3. How could environmental scanning have made a difference?
4. What informal activities could the fire officers have relied on to gain information?
5. What formal activities could the fire officers have pursued to become involved in community planning efforts?
6. Should the mayor have appointed the fire chief to the planning committee? Why?

Note: This could be either a full-time or combination department.

Case #2

The Allen Township Volunteer Fire Department station is located on a busy county highway. The county decides to complete its planned road improvements and sets up orange barrels to begin the process.

During the afternoon of the first day of construction, ATVFD receives an EMS call and the crew discovers that access streets to the section of the township on the other side of the road have been closed off. They are forced to retrace their path and then detour around the construction, causing a delayed response.

Following their return to the station after transporting the patient, the crew leader is furious and calls the fire chief to complain bitterly. The fire chief admits that he had been made aware that some construction would take place this spring; however, he was not aware that access roads would be closed off or that construction would begin this soon.

An officers' meeting is immediately scheduled to discuss the ramifications of the construction project. The township trustee, the road superintendent, and the police chief are invited to explain their roles in this project. Each states that they have made adjustments to account for their normal operations, but had not thought about the impact on the fire department. The construction project had been discussed in detail at the last township trustees' meeting, but the fire chief does not normally attend these meetings and was not present. The minutes of the meeting had been sent to all department heads, including the fire chief. The chief had to assist in responding to an emergency call the day the minutes arrived and simply posted them, intending to read them later, but had unfortunately never found the time to do so.

1. What could the fire chief have done to prevent this situation?
2. Who could have assisted the fire chief?
3. Why didn't the other department heads think about the possible impact on the fire department?
4. What should the chief do about the road construction now?
5. What should the fire chief do to prevent something like this from occurring in the future?

6. What type of informal information could have provided a clue that the road was going to be closed?

Case #3

The Portageville Fire Department is a combination department with four full-time members on duty each shift in addition to a fire prevention officer and the chief, who each work 8-hour shifts. In addition, thirty-four paid-on-call members respond as needed. The community is beginning to experience some growth, as evidenced by several new housing developments.

A new four-story motel is under construction within the fire department's jurisdiction. It is nearly complete. After passing the new building on the way back to the station one day, the fire chief asks the fire prevention officer whether she has seen the plans for the building or made an inspection. The fire prevention officer explains that the county building department approves all plans and, therefore, she has had no access to them. According to the state building code, the new motel must have an automatic sprinkler system installed. However, neither the fire chief nor the fire prevention officer have any knowledge of the fire protection equipment installed or of the layout of the means of egress.

That afternoon they both visit the new motel and are appalled to discover that the siamese connection for the sprinkler system has been located at the rear of the building in a location that is difficult to access. The chief immediately contacts the construction superintendent and demands that the siamese connection be moved to the front. Although the superintendent indicates his desire to cooperate, he explains that the water supply system and riser are already installed and, therefore, could only be changed at great expense to the owner.

1. What course of action should the fire chief take?
2. Why didn't fire department officials know that the motel was going to be constructed prior to ground breaking? How could they have found out?
3. What action could the fire prevention officer have taken prior to this incident?
4. Who should the fire prevention office have contacted?

National Professional Organizations (Fire Service Based)

International Association of Fire Chiefs
1329 18th Street, NW
Washington, DC 20036-6516

International Association of Fire Fighters
1750 New York Avenue, NW
Washington, DC 20006-5395

National Fire Protection Association
1 Batterymarch Park
P.O. Box 9101
Quincy, MA 02269-9101

National Professional Organizations (Not Fire Service Based)

American Management Association
135 W. 50th Street
New York, NY 10020

American Society for Public Administration
1120 G Street, NW, Suite 500
Washington, DC 20005

American Society for Training and Development
1630 Duke Street, Box 1443
Alexandria, VA 22313

International City Management Association
1120 G Street, NW
Washington, DC 20005

National Association of Counties
440 First Street, NW
Washington, DC 20001

National Coordinating Council on Emergency Management
7297 Lee Highway, Suite N
Falls Church, VA 22042

National Emergency Management Association
c/o James Maher
1410 Riverside Drive
Jackson, MS 39202

ENDNOTES

1. J. D. Williams, *Public Administration: The People's Business* (Boston: Little, Brown and Company, 1980), p. xii.
2. Ronny J. Coleman and John A. Granito, eds., *Managing Fire Services* (Washington, DC: International City Management Association, 1988) and Thomas E. Drabek and Gerard J. Hoetmer, *Emergency Management: Principles and Practice for Local Government* (Washington, DC: International City Management Association, 1991).
3. Richard J. Stillman II, *Preface to Public Administration: A Search for Themes and Direction* (New York: St. Martin's Press Inc.), p. 88.
4. Although the fire department may not be an independent unit in some cities or districts where the fire service is but one component of a department of public safety, the overall point of management responsibility is the same. It is important for the reader to consider the substance of this discussion beyond the organizational position of individual fire departments in a given local jurisdiction.
5. Ronny J. Coleman, *Residential Sprinkler Systems: Protecting Life and Property* (Quincy, MA: National Fire Protection Association, 1991), p. 117.
6. Nancy K. Grant, "Forms of Local Government," lecture for Master's course entitled *Introduction to the Profession of Public Administration,* 1982–1993.
7. A more comprehensive discussion of this topic can be found in Nancy K. Grant and David H. Hoover, "Interagency and Intergovernmental Cooperation Requirements for Small Communities Under SARA Title III," in *New Directions in Public Administration Research,* Vol. 2, No. 3, Oct. 1990. Much of the following discussion is based on that article.
8. Barry H. Jordan, "Superfund and Community Right-to-Know: Power to the People," in *Hazardous Materials Waste Management Magazine,* May–June 1987, pp. 22–23.
9. David A. Aaker, "Organizing a Strategic Information Scanning System," *California Management Review,* Vol. 25, Jan., pp. 76–83 passim.
10. John B. Olsen and Douglas C. Eadie, *The Game Plan* (Washington, DC: The Council of State Agencies, 1982), p. 29.
11. Jay S. Mendall, *Nonextrapolative Methods in Business Forecasting* (Westport, CT: Greenwood Press, 1985) and David C. Korten, "Strategic Organization for People Centered Development," in *Public Administration Review,* vol. 4, July–Aug., 1984, pp. 341–352.
12. Olsen and Eadie, *The Game Plan.*

REFERENCES

Aaker, David A. 1983. "Organizing a Strategic Information Scanning System," *California Management Review.* 25: 76–83. January.

Coleman, Ronny. 1991. *Residential Sprinkler Systems.* Quincy, MA: National Fire Protection Association.

Coleman, Ronny J., and John A. Granito. eds., 1988. *Managing Fire Services.* Washington DC: International City Management Association.

Drabek, Thomas E., and Gerard J. Hoetmer. 1991. *Emergency Management: Principles and Practice for Local Government.* Washington DC: International City Management Association.

Jordan, Barry H. 1987. "Superfund and Community Right-to-Know: Power to the People." *Hazardous Materials and Waste Management Magazine.* May–June. pp. 22–23.

Korten, David C. 1984. "Strategic Organization for People Centered Development." *Public Administration Review.* July–August. 4:341–352.

Mendall, Jay S. 1985. *Nonextrapolative Methods in Business Forecasting.* Westport, CT: Greenwood Press.

NFPA 1021, *Standard for Fire Officer Professional Qualifications,* 1992 edition. Quincy, MA: National Fire Protection Association.

Olsen, John B., and Douglas C. Eadie. 1982. *The Game Plan.* Washington DC: The Council of State Agencies.

Stillman II, Richard J. 1990. *Preface to Public Administration: A Search for Themes and Direction.* New York: St. Martin's Press, Inc.

United States Federal Government. 1986. "Public Law 99–499: The Superfund Amendment and Reauthorization Act: Title III—Emergency Planning and Community Right-to-Know Act."

Williams, J. D. 1980. *Public Administration: The People's Business.* Boston, MA: Little, Brown and Company.

CHAPTER 8

Information Management

The 1980s and 1990s have often been called the "information age." The amount of information available and shared on a regular basis is staggering. This surge in information occurred primarily because of technological advances in communication and data processing systems as well as changes in methods of decision making and organizational management. More information in diverse formats can be processed and shared quickly. With the increased availability of diverse information, decision makers have come to demand more information to make more wide-ranging, comprehensive decisions.

Information management means more than simply putting data on the computer!

Information management refers to the system of information collection, processing, dissemination, and use. The computer hardware involved is simply a tool of the entire system. It is important to have an effective information management system within a fire department and to recognize the role the department plays in the larger information system of the fire service. This chapter will address these issues from a systems perspective and will discuss both the administrative and technical aspects of information management.

The need for an information management system in the fire service has been recognized for some time. As the fire service grows more sophisticated and fire administration develops as a profession, the use and importance of information will increase. Fire officers will be called upon to present documented evidence of need and performance. As accountability for performance increases, the need to be able to substantiate performance claims becomes correspondingly important. The use of information as a management tool is expanding at both the local and national levels. As information about fires and fire departments becomes more available, not only are favorable laws and regulations enacted, but local administrators and public officials become more aware of what the fire service does. They then want a greater amount of information from and about their own community's department. Local fire officers must be prepared to respond to these demands.

PROPER USE OF INFORMATION

Information management is crucial to the effective management of any organization. The purpose and use of information should determine the collection format and procedures. In addition, the use of the information should also be the factor that determines how the information is processed and presented. As information manager, it is not necessary for a fire officer to be a statistician, researcher, or computer whiz. Instead, information managers should be familiar with decision making, quality assurance, planning, and budgeting. These four crucial administrative functions rely heavily on data and information. The officer charged with information management must match the data collected, processed, and disseminated with its intended use. Unless the officer understands how the information is going to be used, he or she cannot be certain of how information should be presented—or even what information must be gathered. This is the essence of information management.

Too often, the computer department or systems and programming staff drives the information system. Because of the technical nature of computer operations, many senior officers do not feel comfortable programming and operating computers. However, it is not necessary to have the technical information about a computer to make decisions concerning the type of data to be collected and the content of reports to be produced.

Computer experts have the ability to produce data in almost any format desired, but they do not necessarily *understand* all of the reports they are producing. This is the difference between information *management* and information *processing*. Management refers to selecting the information to be collected and reported. Processing is the collection and analysis of the requested data. Fire officers can benefit from being able to perform both functions. For senior officers, however, it is the management of information that is crucial.

Determining what information is to be collected is sometimes difficult. The National Fire Incident Reporting System (NFIRS) is one system that can be used to facilitate the process. Other departments use run sheets that were designed by their department. Either of these may be appropriate, but in order to evaluate the effectiveness of an information management system, it is better to be driven not by the form, but by the information needed.

In evaluating the public safety services (police and fire) of a small city, the authors recently met a police chief who was in charge of the data processing operations. He proudly challenged us with the statement: "Ask me anything you want about our responses. If you can ask a question I can't answer, I'll be surprised and then thank you because that will indicate a hole in the system we haven't plugged yet." Although a bit extreme, this statement illustrates how one police chief measured the success of his information system: the ability to answer questions from administrators and evaluators!

The key to good information management is knowing why information is being collected and how it is being used. In order to identify this, it is necessary to understand: (1) operations, (2) information uses and needs, and (3) information systems.[1]

USE OF INFORMATION BEYOND THE FIRE DEPARTMENT

Information gathered by local fire departments is also used on a state and national basis to assist the development of the fire service in general. This can only be accomplished with the assistance of local departments gathering and forwarding the requested information. The National Fire Incident Reporting System (NFIRS), which is operated by the U.S. Fire Administration, is the single largest data collection maintained on a national level. It is used by various fire departments and fire service organizations on an ongoing basis. Understanding the background and use of NFIRS demonstrates how it benefits local fire departments.

The National Fire Incident Reporting System (NFIRS)

NFIRS is best described as "a tool for fire departments to report and maintain computerized records of fires and other fire department activities in a uniform manner." As is widely recognized, the report issued by the President's Commission on Fire Prevention and Control in 1972, "America Burning," was a critical turning point in the development of the modern fire service. In addition to bringing widespread recognition to the fire problem and the service provided by fire departments across the nation, it also documented a need for more accurate information about fire response operations. In reaction to this report, Congress passed Public Law 93-498, The National Fire Prevention and Control Act, which established the National Fire Prevention and Control Administration, the forerunner of today's U. S. Fire Administration. Among other activities, the agency was charged with the creation of a data collection project, known today as NFIRS.

In building this database, the Fire Administration utilized the groundwork laid by the National Fire Protection Association and adopted its standard NFPA 901, *Uniform Coding for Fire Protection.* This dictionary of fire terminology was the first attempt to standardize and codify terms used to identify types of incidents, injuries, and response. This standard is periodically reviewed and updated. There are four major objectives of NFPA 901:

1. To provide for the collection of information required for legal record purposes and control of the fire problem
2. To provide local fire service management with information to indicate trends; to measure the effectiveness of fire prevention, fire suppression, and emergency mitigation procedures presently in practice; to evaluate the impact of new methods; and to indicate those areas that may require further action
3. To provide a pre-fire inventory of property in a fire service district so that future needs may be anticipated and potential problems corrected before a fire

Incident report ______________________ Fire Department

NFIRS 1

1 ☐ Delete
2 ☐ Change

Fill in this report in your own words

Complete for all incidents

A FDID | Incident # | Exp # | Mo | Day | Year | Day of wk | Alarm time | Arrival time | Time in service

B Type of situation found | Type of action taken | Mutual aid 1 ☐ Rec'd 2 ☐ Given

C Fixed property use | Ignition factor

D Correct address | Zip code | Census tract

E Occupant name (Last, first, mi) | Telephone | Room or apt.

F Owner name (Last, first, mi) | Address | Telephone

G Method of alarm from public | District | Shift | No. alarms

H Number fire service personnel responded | No. engines responded | No. aerial apparatus responded | No. other vehicles responded

Complete if casuality

I Number of injuries Fire service | Other | Number of fatalities Fire service | Other

Complete for all fires

J Complex | Mobile property type

K Area of fire origin | Equipment involved in ignition

L Form of heat of ignition | Type of material ignited | Form of material ignited

M Method of extinguishment | Level of fire origin | Estimated loss (dollars only)

Complete if structure fire

N Number of stories | Construction type

O Extent of flame damage | Extent of smoke damage

P Detector performance | Sprinkler performance

Q If smoke spread beyond room of origin | Type of material generating most smoke | Avenue of smoke travel

R Form of material generating most smoke

S If mobile property | Year

T If equipment involved in ignition | Year

☐ Check if comments on reverse side

U Officer in charge (name, position, assignment) | Date

V Member making report (if different from above) | Date

FIGURE 8-1 The National Fire Incident Reporting System Data Form

4. To provide uniform data to regional, national, and international fire and emergency organizations in order: to make the full extent of the fire and emergency problem known; to reveal facts that require action on these levels; to guide the effective development and administration of codes and standards; and to guide fire prevention, fire protection, emergency medical treatment, and hazardous materials handling research.[3]

This is the first attempt to utilize uniform coding and definitions of fire incidents in order to develop a comparable database.

More than forty states and more than 14,000 fire departments participate in NFIRS, resulting in a database of more than 500,000 incidents annually. Local communities participating in NFIRS can compare their trends with those on a national and state level. Some cities are using the data for focusing the subject of fire prevention and education programs.

States have used NFIRS to substantiate the need for legislation such as sprinkler installation and control of the sale and use of fireworks. On the national level, the National Fire Data Base is used to compile a number of annual reports. It is used by both private organizations and government agencies. Both the NFPA and the Consumer Product Safety Commission have used NFIRS to improve the level of fire safety in the U.S.

Uses of Data by National Organizations

A number of national organizations and federal agencies deal with various topics of interest to the fire service. This section provides a brief overview of some of the more prominent ones.

The NFPA conducts an annual survey that supplements the data gathered on the NFIRS forms. This survey provides a profile of the changes in the fire service and in experiences with fires. While NFIRS collects information on each incident, it does not look at the composition of the fire department, changes in staffing, or types of equipment available. These areas are considered in the supplemental survey.

The information gathered is used to support the NFPA standards-making process and to support national and local legislation to improve the level of fire safety. This use of data to advance the safety and effectiveness of the fire service is here to stay.

> *Fire protection is becoming increasingly scientific. The advent of sophisticated new data bases, measurement techniques, and computer-based models in just the 1980s has produced a pace of change unlike anything seen before. . . .*
>
> *There is an increasing demand for hard evidence of the effectiveness and the cost-effectiveness of fire protection features, systems, codes, and standards.*[4]

As budgets continue to tighten and the demand for accountability increases, the need to document and verify the need for services and the services themselves will remain a necessity.

The Consumer Product Safety Commission has used information from the NFIRS reports to identify appliances and other consumer products that may be potential fire hazards. When the national aggregation of data from the NFIRS document a significant number of fire incidents caused by a specific appliance, the Commission researches that trend for verification and then conducts laboratory tests to determine whether the product does, indeed, pose a fire hazard to the public. If so, the Commission can then impose requirements for recalls and warnings to consumers. In addition, the Consumer Product Safety Commission supported the installation of smoke detectors and conducted several tests that verified their effectiveness in detecting a fire and sounding an alarm.

Thus, national organizations have produced and analyzed data relating to fire incidents at the local level in order to increase public safety. Fire departments that have experienced a reduction in the number of fires and in the number of injuries and deaths due to fire have benefited directly from the use of such information.

Other Uses and Sources of Fire Data at the National Level

- The Building and Fire Research Laboratory (BFRL), a division of the National Institute of Standards and Technology in the U.S. Department of Commerce, conducts large-scale fire tests and produces technical publications detailing the results.

- NFPA's Fire Incident Data Organization System (FIDO) provides detailed information on fires deemed to be of high technical interest.
- The International Association of Fire Fighters (IAFF) releases annual totals of on-duty fatalities and injuries of member fire fighters.
- The National Institute for Occupational Safety and Health (NIOSH) studies injuries due to fire in various occupations. Starting in 1972, the National Electronic Injury Surveillance System (NEISS) has tracked product-related injuries in homes, including those from burns and electrical shocks due to causes other than fire.
- The American Burn Association tracks burn patients better than the National Health Inventory survey of the U.S. Department of Health.
- The National Transportation Safety Board (NTSB) publishes reports on aircraft, railroad, and hazardous materials highway accidents. Fatal highway accidents are tracked by the National Highway Traffic Safety Administration (NHTSA).
- The U.S. Coast Guard compiles information on incidents involving recreational boats and commercial vessels, while the Naval Safety Center collects incident records from all of the military services.
- The International Association of Electrical Inspectors (IAEI) and Underwriters Laboratories, Inc. (UL) maintain a clipping file of published information concerning electrical fires and shocks.
- The U.S. Forest Service prepares an annual report detailing fires that have occurred in national, state, and private forests.
- The Department of the Interior Bureau of Land Management reports on fire incidents on land it owns and manages.

Thus, when local fire departments gather and report fire statistics, they have a major influence on national decisions relating to fire prevention. These national responses, in turn, influence local fire departments as legislation is amended. Individual departments and officers can utilize the results of this research to better train and prepare for incidents, to support more effective building and life safety codes, to substantiate the need for local ordinances, and to define public education programs.

Benefit to the Local Department

National data information and reports are valuable resources for local fire officers and departments. For example, when trying to implement local sprinkler ordinances, national data sets can be utilized to document their need and effectiveness. Ronny J. Coleman advises:

> *A good way to begin is to look at the fire problem on several levels. First, leading proponents of a residential sprinkler system must become conversant with what makes up the nation's fire problem. This includes learning relevant and convincing fire data and statistics, as well as developing an understanding of the fire problem at the local level.*
>
> *Several documents, such as America Burning and the NFPA Annual Fire Loss report, help form the most fundamental arguments in favor of the life safety sprinkler system concept. . . .*
>
> *Residential sprinkler proponents must develop an in-depth understanding of the local fire conditions that reflect state and national trends. Consequently, some of the most important research documents in support of a residential sprinkler ordinance are the fire records of the local community.*[5]

Thus, the national data reports provide a basis of comparison with which local fire departments can compare their own performance. State-level information and data reports enable the same type of comparison. This provides important direction to a department that has an unusually high number of incidents of a certain type. For example, communities with a larger than average number of mobile home fires may need to consider specialized SOPs, increased voluntary residential fire inspections, focused public education efforts, or more effective building and zoning ordinances. This type of information can be very beneficial to the increased fire safety of a community. However, its use is only possible through the collection and analysis of national information *and* the use of local fire department information bases. The fire department must first document the number of mobile home incidents it is experiencing and then compare this with state and national trends.

The previous discussion examines some ways in which national information processing systems have assisted the development of the fire service and the increased safety in local fire districts. The remaining sections of this chapter will examine the use of information systems within a local fire department. Remember, the most important indicator of whether a fire department is doing a good job of information management is how well the information is *used.*

A SYSTEMS VIEW OF FIRE RESPONSE

A systems view of fire department operations demonstrates the interactive components of information management in constant use. Department operations are dynamic, constantly changing to some extent, requiring fine-tuning and amended decisions as changes occur in the department and the community.

In order to respond to a call, a fire department needs personnel and equipment (including apparatus). These are the two primary "raw material" inputs into the fire department system. Before entry, however, equipment and personnel must be screened to ensure that they are of appropriate quality and standard for intended use. Thus, the equipment must be evaluated according to established criteria. These criteria may be found in information sources such as NFPA 1500, *Standard on Fire Department Occupational Safety and Health Program,*[6] and department bid specifications. The department officers making purchasing decisions must have access to the appropriate information.

Personnel entering the fire service must also be screened. Full-time fire departments gather various pieces of information when deciding whether to hire a potential fire fighter. These include civil service test scores, physical agility results, psychological test results, and results from physical exams. Even volunteer fire departments usually have some type of screening requirements. Many states mandate physical examination results to be on file prior to hiring volunteer or part-paid fire personnel. This becomes part of the information system—the baseline physical data on each new department member. In addition, many departments consider some of the following: driving records, felony records, age, education level, and residency. Regardless of specific requirements, the common factor is that some type of information base is being used to add each individual to the staff. This is the beginning of the information system. So far there are two primary information groups: personnel and equipment.

The new equipment and personnel must then be transformed into "usable components" of the response system. The apparatus and equipment are appropriately set up and then tested and maintained on an ongoing basis. Personnel are trained and tested (or evaluated in some manner) to ensure they are capable of performing the required duties in a response situation. This is done to achieve the intermediary goal of "preparedness." Again, a certain set of information is added to the information system. The information system is now growing according to use. The department can adequately document its degree of preparedness by delivering information on its equipment and personnel.

The next external impetus is an incident or call for assistance. At this point, the fire department enters response mode, which is its primary function. In order to determine whether the response is appropriate (an evaluative question that will be asked for purposes of quality assurance and continued operation support), certain information must be gathered and monitored. This is where run reports enter the system. (This discussion will focus on the information relevant to the two department components used in this illustration.) Monitoring which equipment was used, the performance of that equipment, and subsequent cleanup, inspection, and repair of the equipment is necessary in order to judge whether the level of preparedness was appropriate and whether changes should be made following the incident. (If equipment or apparatus is damaged or lost during an incident, the level of preparedness is decreased until that equipment is either repaired or replaced.) The time of response and the performance of fire fighters responding are important measures of adequate preparedness. Time of response is almost automatically gathered. Performance is another issue, but debriefing can assist in gathering this information. Another important aspect for volunteer and part-paid departments is the identity of responders and what position they held. This can be useful information for full-time departments, too, as it tracks the amount of experience a fire fighter gains. In addition, on a volunteer or combination department, this is an indicator of how many and which fire fighters can be expected to respond to an incident. Thus, in terms of prepared-

FIGURE 8-2 A Systems Explanation of Fire Department Operations and Information Flow

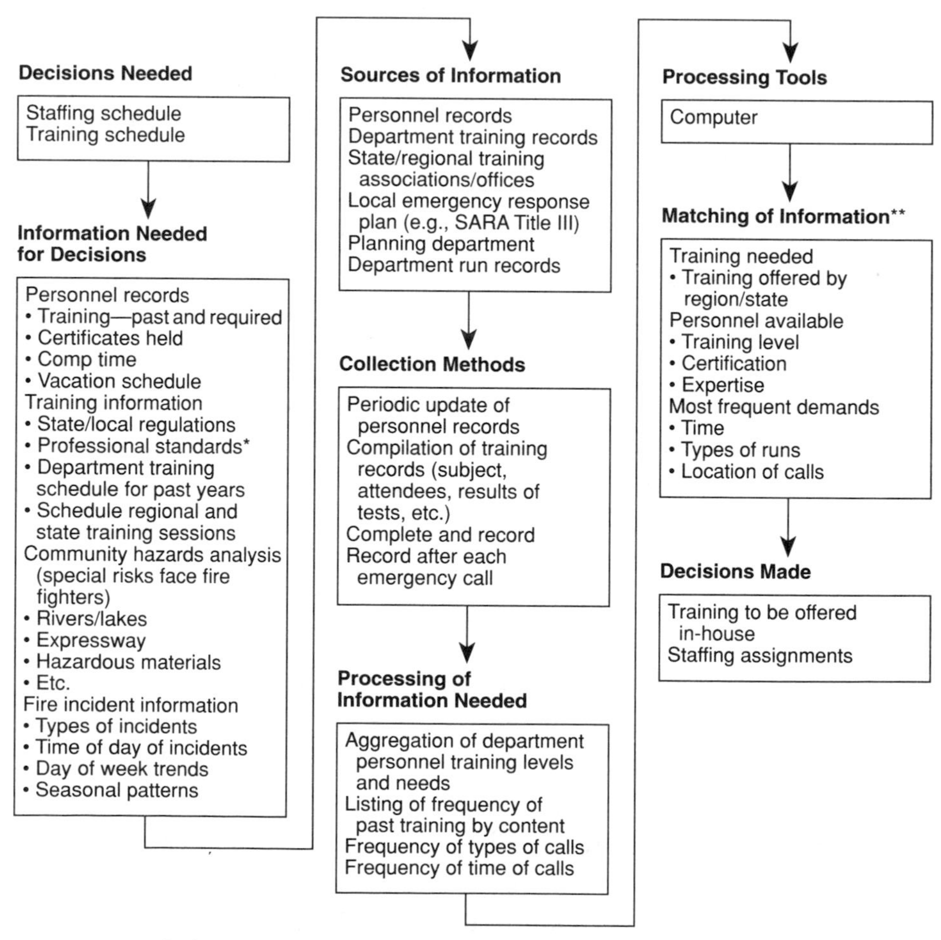

*NFPA 1500, *Standard on Fire Department Occupational Safety and Health Program*

**Similar to gap analysis described in Chapter 10

ness, it can indicate whether more personnel need to be added to the department roster.

Information about these activities is then used to identify what type of training is needed, how well equipment is standing up to use, which fire fighters demonstrate aptitude in different areas, the level of service provided, and what the cost of continued or enhanced service will be. Thus, information is used on an ongoing basis. It is this use that should drive the collection, processing, and reporting of data and information.

Figure 8-2 demonstrates how the decisions to be made drive the development of the information system. Note that the computer is only a small part of the overall process. It is a tool used to process the data and information according to the needs of the decision makers. This chart does not show all of the minute information involved in the process, but it does serve to illustrate how an information system is built from the basis of use and decision.

LOCAL BENEFITS OF A FIRE REPORTING SYSTEM

There are many uses for a fire reporting system at the local level. Regardless of the size or type of department—paid, combination, or volunteer—good information about operations can be used to plan for the future, demonstrate accomplishments of the past, and identify ongoing activities. NFPA 901, *Uniform Coding for Fire Protection,* provides standard definitions that enable all collection of fire department data to be comparable. Using its definitions in a local reporting system enables the department to not only use the information in a number of activities, but to compare their activities with those of other departments and even national trends. The following is a brief overview of a number of different uses of data by local fire departments.

- *To describe and document the community's fire problem.* A graphic and accurate presentation of exactly what type of fires have occurred in the past can be drawn from a report presenting the fire trends of the past few years. Figure 8-3 shows trends in types

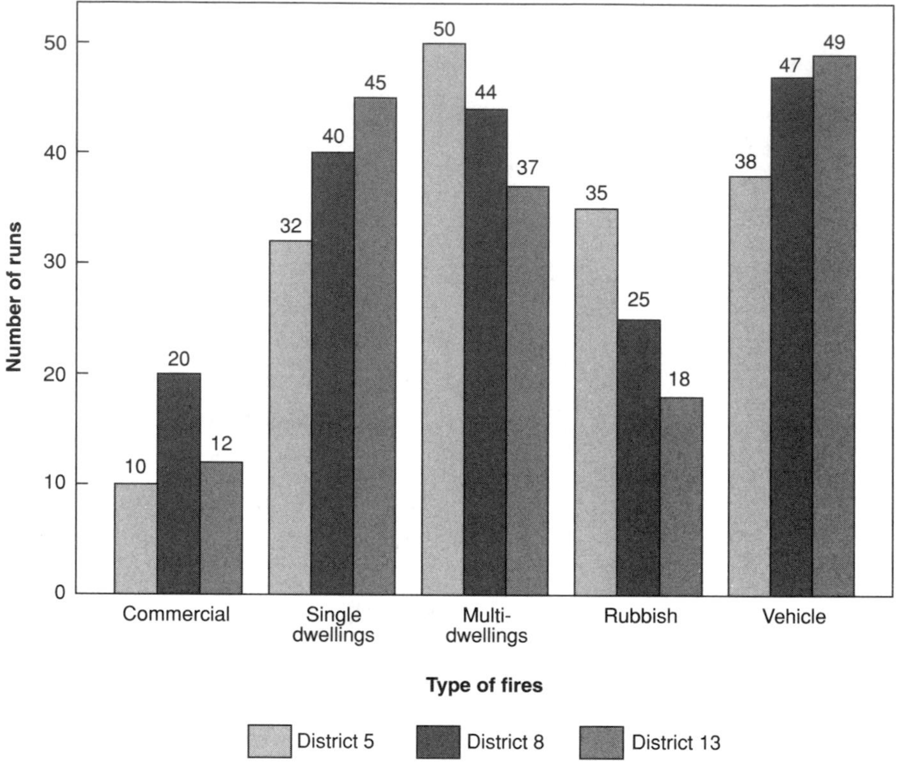

FIGURE 8-3 Trends in Fire Runs, 1990, 1991, 1992

of fires in a community with targeted public education. What topics were targeted?

- *To plan and set priorities for future fire protection needs on the basis of assessed needs.* Once the local fire problem is clearly defined, the relative degree of different problems and the potential contribution of various prevention and suppression preparedness activities can take place. In this manner, resources can be allocated first to the most serious aspects of the local fire problem.
- *To support budget requests.* Local department presentations of trends and activities can be used to substantiate requests for increased resources, or even to defend maintenance of current resource allocation. Without this hard data and information to substantiate the level of service provided and the need for the fire department, it is increasingly difficult to convince public

administrators and local officials to continue funding at existing levels and to increase funding.

- *To demonstrate the need for local ordinances.* Data can be used to demonstrate the need for more stringent local building ordinances. Trend analysis of department fire calls and comparison with areas having the desired ordinances can clearly show the effect that a new ordinance would have on reducing fire risk.
- *To evaluate fire inspection and code enforcement programs.* Data on fire inspection programs coupled with information relating to the cause and origin of fires in a community can be used to evaluate the impact of local code enforcement efforts. Having a code in place and writing an inspection program is a start, but if it is not carried out, it is almost useless. Analysis of annual fire statistics can reveal whether a program is operating as planned and how effective it is.
- *To evaluate public education programs.* Information on the cause of fires can be used to identify needed areas of public education. In addition, tracking these same causes following a specifically targeted public education program can then demonstrate the effectiveness of the program. For example, a community experiencing a rash of grass fires in the summer discovered that a majority of the fires were caused either by fireworks or matches from groups of individuals lighting fireworks. A concentrated public information campaign was started in the fall and expanded in the late spring and early summer. Data from the following year demonstrated that the number of grass fires had fallen not only below the number for the previous year, but also below the average for 5 years prior to that.
- *To schedule nonemergency activities.* Information on the run patterns in the community enables officers to schedule nonemergency activities such as training and equipment maintenance during off-peak hours. This facilitates the completion of these activities while maintaining the highest possible level of preparedness.
- *To schedule staffing.* Especially in volunteer and combination departments, it may not be possible to have the same number

of people on standby duty or even on-call all the time. An analysis of run patterns allows fire officers to schedule more personnel on standby and on-call when there is the highest risk of an incident. This type of scheduling may also be used by combination and full-time departments when severe budget restrictions prohibit full staffing at all times.

- *To demonstrate the need to upgrade the level of service.* Changes in the local community may create a need to upgrade existing

TABLE 8-1 Fire Calls—1990

1990	*Mon.*	*Tues.*	*Wed.*	*Thurs.*	*Fri.*	*Sat.*	*Sun.*	*Total Calls*	*% per 4 hr*	*% per 6 hr*	*% per 12 hr*
0001-0100	1	1		1	1			4			
0101-0200						1		1			
0201-0300				1	2			3	9.6	12.0	
0301-0400	1	1		1		1		4			
0401-0500								0			
0501-0600			1		1	1		3	4.8		33.6
0601-0700	1							1			
0701-0800		1			1			2			
0801-0900			2		1	2	1	6			
0901-1000	1	1	2		1		1	6	19.2	21.6	
1001-1100	1	2	1	1		1		6			
1101-1200	1	1	2	1	1			6			
1201-1300	1						1	2			
1301-1400		2	1	2			1	6	16.8		
1401-1500	1	1		1	1			4		32.0	
1501-1600	2	1	2	2	1	1		9			
1601-1700	1			2	1	2	1	7			
1701-1800	1	2	4	1		3	1	12	28.8		66.4
1801-1900	1	2	3	1	2	1	1	11			
1901-2000	1	2		1	1		1	6			
2001-2100			3	1				4			
2101-2200	1	3	1	2	5		1	13		34.4	
2201-2300						2	3	5	20.8		
2301-2400			2	1		1		4			
TOTAL CALLS	15	20	24	19	19	16	12	125			
Percentage	12.0	16.0	19.2	15.2	15.2	12.8	9.6				

service levels of the fire department. Communities offering Basic Life Support (BLS) Emergency Medical Service (EMS) in aging communities may well experience an increase in the number of EMS calls that require advanced life support. Documenting this by means of a tally and trend analysis of types of EMS calls may substantiate the need for upgrading the level of service. Similarly, communities with an increase in commercial and industrial structures can use information from the building department to document the need to increase staffing in the fire inspection program or expand activities of fire inspectors to enable annual inspections of all existing industrial and commercial structures.

Fire departments can utilize information in a variety of ways. The key is to (1) have accurate information, (2) have the data in a form that can be used, and (3) relate the information appropriately.

FIRE DEPARTMENT REPORTS

Fire department reports can take many forms, ranging from the individual run sheet to the annual report presented to council. Basically, a report is a document containing information that is shared for a specific purpose. The purpose should determine its design. Some reports are used for more than one purpose, such as runs sheets, which report what happened on a given call and also serve as a data collection sheet for more comprehensive reports. When a report serves multiple functions, all uses of the information should be considered when deciding what information to collect.

Designing and Selecting Data Collection Forms

Most fire departments use run sheet reports. Many departments use the NFIRS report forms as their own run sheet; others have a department run sheet and also complete the NFIRS forms (see Figure 8-1). The NFIRS report form contains most of the standard information about the response to an incident and the nature of the incident.

Some departments, however, especially volunteer and combination departments, need additional information on their daily operations. For example, a department may need to monitor who is responding to calls and which apparatus are used to respond.

Criteria of Good Data Collection Forms

Whether adopting an existing form or creating one specifically for one's department, there are a number of things to consider:

1. What data are legally required to be collected?
2. What information is this system expected to produce and for what purpose?
3. What data elements are necessary to provide the information needed?
4. How quickly after the incident is the information needed, by whom, and how long after the incident should the information be kept available?
5. Who will be expected to collect the data, assemble it, and screen it for errors?
6. Who will be expected to summarize and analyze the data?
7. What savings or efficiencies could be realized if the data were available?
8. Is any of the needed data already being collected for another use?[7]

Make sure the form contains all of the information necessary. Although categories of information can be added to later versions of the form, it is almost impossible to go back and add the information to those records that have already been completed. The result is a group of initial data records that are incomplete. This is one reason why pilot testing a new form is advised.

Make the form as simple as possible. Reduce the amount of writing required to a minimum and make certain the categories are clear. The more complex the form, the more likely errors will occur during completion.

Put all entries in logical order. The sequence of data collection should flow in a logical format. For example, forms on personnel preparedness might include a section dealing with formal certification. A logical progression from this would be to then add a section detailing the type of specialized training completed. This would make more sense than to jump from certification to demographic information, such as age, gender, race, place of residence, and then back to specialized training and formal education.

Make sure the instructions for completing the form are accurate and clear. Each form should be accompanied by an instruction guide that explains how to answer each question.

Train ***everyone*** *who may be asked to complete the form.* An instruction guide is not a substitute for training. Showing staff how to complete a form and making them familiar with the instructions is critical to accurate data collection. Orientation for each new member of the department should include an introduction to the information processing system with detailed instruction covering those aspects of the system with which he or she may have contact.

If the information will be entered into a computer system or compiled for an aggregate report, create a code book detailing how the information is to be set up in the file. With the increased sophistication of computers, many of these code books are simply programmed into the computer system and the data are entered directly from the data collection forms to the computer database. This is the most labor-efficient method, once it is in operation. Every system should be thoroughly tested before going on-line. It is normal to find flaws in a new system or database. This is why time should be scheduled prior to start-up in order to use and test the system. Any adjustments can be made at that time.

RECORD KEEPING

Regardless of what type of records are being kept, training everyone who uses them is crucial. One group of people may be filling out the forms while another group is entering data on the computer system, and yet a third is requesting the analysis for decisions. It is vital that each of these groups share the same understanding of definitions and codes used in the records.

Security of records is another critical issue. This is especially true of personnel records, which should be accessed only by authorized individuals. Other records also must be secured in a manner that prevents them from being altered or lost. On computer systems this can be accomplished through the use of passwords and limited-access locks. In departments where records are kept and compiled by hand, they can be secured in a locked file cabinet. This security is important in order to verify that the information used in reports is accurate.

Another aspect of ensuring accuracy is the quality assurance system designed by the department to ensure that the data is not entered erroneously or accidentally changed during the process. Quality assurance should utilize random spot checks of records to ensure that the entry into the computer system matches the hard copy on the completed forms. In addition, any glaring exceptions to the norm should be double-checked against the hard copies for accuracy. If they match, it may then be necessary to interview the individual who completed the form in order to discover the details of the exceptional situation reported.

For example, in a volunteer department with an average response time of 7.8 minutes when no standby crew is at the station, the average response time over a particular month for this situation was 25 minutes. Of the three such runs recorded that month, one showed a response time of 32 minutes, while the other two were under 6 minutes. Checking the hard copy revealed that the time of alarm on the questionable department run sheet was 10 minutes later than that in the computer record. The time in the computer record was the time at which the police department was dispatched to the scene—the first 911 call, not the time at which the fire department EMS unit was dispatched. (The incident was a murder-suicide.) Interviewing the paramedic who responded to the call, the fire chief discovered that the ambulance was detained one block from the scene by the police department since they had not yet secured the scene and found the perpetrator. This resulted in another 10-minute delay in responding. Thus, the actual response time for the call was 12 minutes, still above average, but not excessive. A special circumstance such as this can easily skew the report results. Thus, a quality assurance system, in addition to ensuring accuracy, can aid in identifying exceptions that need to be noted rather than simply included in the totals.

One of the most fundamental aspects of record keeping is consistency. The same information must be kept for each record. In other words, if the type of apparatus is listed on one report form, it must also be on another. If the personnel files of one fire fighter make note of formal education accomplishments, the records of each fire fighter must show formal education accomplishments. This is especially important in comparative data and report situations. Obviously, when comparing candidates for promotion or special assignment it is important to have the same information on each eligible candidate. In compiling trend reports it is also important to have information in a format that can be used to present annual trends and comparisons.

Types of Department Records

There are a number of records that should be kept by every fire department. The exact content and format of these records may vary, but certain basic information should be contained in each. The following are some of the most vital records to keep.

Personnel records must be kept for each member of the fire department, both volunteer and full-time. These records will cover the individual's service from the time of first application through the date of retirement. They will include: performance reviews at the end of probation, annual evaluations or performance reviews, training completed, transfers, special assignments, certification, commendation, promotions, and disciplinary actions. In addition, it is increasingly important to include baseline as well as continuing medical examination information. This is especially critical when any member of the department is involved in the mitigation of a hazardous materials incident.

In addition to the above information on the individual, records of training and service accomplishments should also be maintained. NFPA 1401, *Recommended Practice for Fire Service Training Reports and Records,* details the type of information that should be contained in these training records. This information can be extremely important in a workers' compensation claim or in cases of a civil suit against the fire department. Lack of information in these areas can also, in extreme situations, lead to liability and lawsuits.

Personnel records covering means of promotion, promotion criteria, testing procedures, and results of testing are also critical. Given today's emphasis on equal employment opportunity and affirmative action, coupled with the changing demographics of American society, fire departments must be able to document adherence to federal mandates and selection on the basis of ability and merit. Inadequate records may prove to be a major liability in challenges to hiring and promotion decisions.

Fire run data summarize the run reports discussed earlier. These records are crucial to the daily operation and future planning efforts of every local fire department, no matter how active or how large. It is vital to be able to document activity in a valid, reliable, and consistent manner. Documented information has a variety of uses and is the key to a progressive department.

EMS run data is comparable to that of fire runs, but extra information needs to be compiled as well. Much of this is dictated by the medical supervisors of the EMS system. It will, at a minimum, include information on patient care administered, status of the patient, diagnosis, and monitoring of status. In addition, training records concerning the completion of continuing education required for maintaining certification levels is mandatory. These vary by state, but most states have continuing education mandates in place for emergency medical certification. A separate set of records need to be maintained on medical equipment (non-apparatus), documenting the monitoring procedures and findings, the testing procedures, and all regular as well as as-needed maintenance. This is crucial to demonstrating continued preparedness for the conduct of assigned responsibilities.

Apparatus inspection and maintenance records should be compiled and monitored. This enables a department to track the status of major apparatus and plan effectively for replacement as needed. It also enables fire officers to monitor apparatus use to determine what the optimal replacement would be, given changes in the composition of the community and corresponding shifts of use of equipment. (Should the second aerial be replaced by a 2000-gpm pumper?) Failure of an apparatus during an incident could then be tracked to determine whether it was due to improper or lack of regular inspection and maintenance. (If so, the department might be liable for negligence!)

Code enforcement fire inspection records should be maintained on each inspection performed in the community and should detail the time of the inspection, who performed the inspection, any violations found, what citations were issued, and what follow-up actions were taken. In addition, the department needs to document how information gathered during code enforcement inspections is subsequently used in developing a preplan emergency response to a specific structure. This last item is critical, as too many fire departments conduct inspections and then file away the information gathered instead of using it to become better prepared to mitigate a potential incident. Other fire departments have failed to follow up on violations. These violations have occasionally resulted in severe incidents, making the fire department look as though it is not doing its job.

These are some of the major records fire departments should keep. Obviously, fire departments must keep records on any program that is mandated by the federal state or local community. For example, federal law mandates reporting of hazardous materials incidents, and fire departments in states and communities with sprinkler ordinances must maintain those records of inspection.

COMPUTER SYSTEMS

During the 1970s and prior, the computers used in the fire service were predominantly mainframes—computers typified by large physical size, significant support requirements (technical staff), and great processing power (for their time in history). The face of computing in government service changed dramatically with the introduction of microcomputers in the early 1980s. This innovation brought with it the potential to move computing power from the "data processing" organization to the end user. The 1990s brought even greater power, and this trend is expected to continue. According to R. M. Patterson, Jr., a consultant who specializes in the information needs of public safety agencies: "By 2000 we will probably have seen the demise of the main-

FIGURE 8-4 Computers are valuable tools for organizing and analyzing data as part of the fire department's information management system.

frame computer for all but very specialized applications. Indeed, the down-sized systems of today have the power of the mainframes of just ten years ago. The information technology market has changed dramatically in just the last several years."[8]

As this book goes to press, the once accepted dichotomy between "large" and "small" computers has become increasingly fuzzy. Local area networks (LANs) have become commonplace and have begun to replace the mainframes and large "mini" computers that once served governments' needs. "Client-server" applications provide for information sharing to a degree not experienced previously. These systems distribute the user-assigned workload between the server, or "host," and the user's computer, or "workstation." The principal effects of this technology have been to drive the cost of hardware (the equipment itself, as opposed to the "programs" or software) down, to decrease the physical size of the systems, to reduce power and cooling requirements, and most importantly, to increase the utility of the systems for the users.

Thus, even small volunteer and combination fire departments are able to purchase and use computer systems for daily operations.

The utility of computer systems increases as the users become more involved with the systems—defining their purpose and determining how the systems should display information. According to Dr. Patterson, users *must* play an active role in determining not only whether computers are needed and what they are to do, but also *how they want to interact with them.* The opportunities for effective integration of computers into the fire service are excellent if the users play an active role in the development of these systems. The users of the system include not only those who will use the outcome information for making decisions, but also those individuals in the department who will be involved in the collection, coding, and entering of data as part of their job responsibilities.

As systems have become less complex to operate, and software has become more "user friendly," the need for highly trained specialists just to *operate* the systems has all but disappeared. New skills—"computer literacy"—are required of more people in the organization. Clearly, this change in technology implies new responsibilities, both for the users of the technology and those who supervise and manage the department's human resources.

The successful users of today and tomorrow will follow a carefully thought-out approach to determining how computers will serve their fire department. They will begin with a systems analysis approach that defines software performance in terms of outputs, and outputs in terms of their own requirements. They will endeavor to find a partnership between software and hardware rather than give primacy to one or the other.[9]

So, how does one *really* acquire a computer system? Certainly one sound strategy may be to take a task force approach. The department's head may select a working group of individuals who have a basic understanding of computing and who also understand their jobs—and the context of their jobs—very well. They should be able to see the "big picture." The task force may be augmented with a member or two from outside the department. (The authors, for example, have had great success with drawing resources from the community served.) Perhaps a consultant can be engaged to help the agency through the process. Employing a consultant may add to the cost of the project initially, but a good consulting firm can help provide assurances that the system will "work" and will often save the department much more than the cost of consulting services. Using experts in system design often makes sense when the project is complex or the department has few resources to use for developing the system.

Those who will determine the desired functionality in the system may try to answer several questions. What system can be acquired that will:

- Improve the effectiveness of operations
- Improve the efficiency of operations
- Increase accountability
- Decrease costs of resources
- Increase management's capacity to understand issues and problems, and formulate solutions
- Help the department's administration rethink operational or administrative problems and see them in a "different light"?

A single computer system that meets all or most of these objectives is a demanding requirement, but it is not unrealistic. While multiple software packages will typically be required to meet such broad objectives, the computer technology of the 1990s permits these packages not only to reside on the same "platform," or hardware, but also allows the interchange of data between them—if these requirements are identified from the beginning.

Ultimately, the right solution will be one that possesses several important characteristics. The "good" computer system will:

- Be a system that includes a mix of software and hardware that does what the users want it to do
- Include hardware that makes sense in terms of compatibility with the software, ease of use, expandability, and versatility
- Be supported by the firm(s) from which it was obtained
- Be cost-justified

Only if these criteria are met will the fire department fully benefit from the integration of computer systems into its operations.

INFORMATION PRESENTATION

Presentation of information is only slightly less important than having the information in the first place. The impact and importance of factual information is too often lost because it is presented poorly.[10]

The first rule for good presentation of data is to make certain the data are clearly understood. Someone with little knowledge of data analysis and only minor knowledge of the fire service must be able to understand the point illustrated by the data. The presentation of data must serve to enhance and clarify the issue, not confuse it. The three most frequent errors made when presenting data are:

- Data are unclear; poor presentation
- Data presented are irrelevant to the subject under discussion
- Too much data are presented and the point is lost

When the data are unclear, the audience often becomes confused. This leads to more questions and challenges instead of resolving and clarifying an issue.

For example, at the fire chief's request, the city council approves the hiring of two full-time fire safety inspectors to reduce the number of commercial fires by 15 percent over the next 2 years. When the city council asks for a report concerning the effectiveness of the program 2 years later, the chief presents three pie charts showing the percentage distribution of types of fires for each year and an annual listing of inspections made. This does not illustrate the program's effectiveness to the council. The presentation should have compared changes in the number of fires by type over the past 3 years with a summation of numbers of inspections by type, violations corrected, and buildings preplanned. This would clearly demonstrate the effectiveness of the additional two inspectors in reducing the risk of fire. The trend analysis of the fire data would show the impact of the program.

The use of irrelevant data occurs when someone presents a lot of accurate data and information, possibly even in clear and under-

FIGURE 8-5 Division of Fire

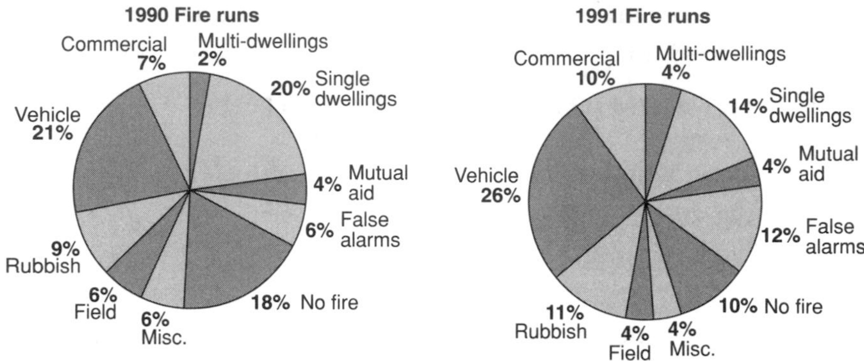

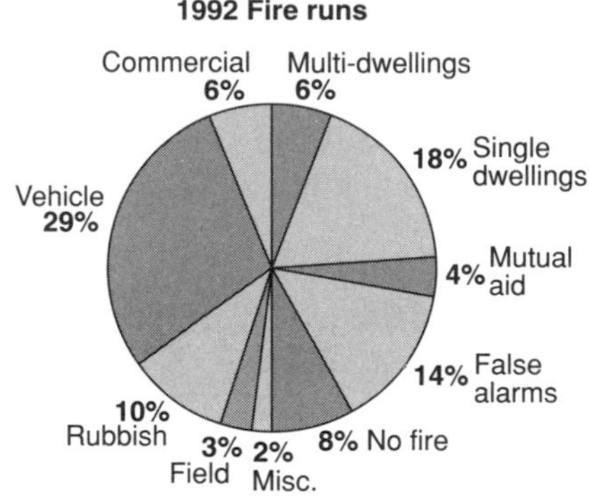

	1990	*1991*	*1992*
Example: steady decrease in number of commercial fires			
Total fires	150	100	135
Commercial fires	11	10	8
% Commercial	7%	10%	6%
		or	
Example: steady increase in commercial fires			
Total fires	289	254	433
Commercial fires	20	25	26
% Commercial	7%	10%	6%
		or	
Example: no set pattern			
Total fires	172	181	184
Commercial fires	12	18	11
% Commercial	7%	10%	6%

Note: Without knowing the number of fires, it is impossible to tell whether there was a decrease or increase of commercial fires in 1991 and 1992.

standable format, that have nothing to do with the question being asked.

The advent of the computer created a data surplus. Many individuals are so thrilled to have data available that they want to share as much as possible whenever possible. As a result, public administrators are so overwhelmed with data that they lose sight of the point being made. For example, the chairman of a community safety committee asks for a report on the accountability and effectiveness of the fire department as one component of a community emergency response capability analysis. The chief presents the following information in response to the request:

- Number of response vehicles
- Miles traveled by each vehicle during the last year
- Gallons of fuel consumed
- Gallons of water pumped
- Total number of feet of hose owned by the department and the condition of each section
- Number of inspections made by the department
- Roster of department members
- Number of hours worked by each member the previous year
- Certification and training record of each member
- An inventory of all hand tools and their condition
- The number of responses for the previous year

Overwhelmed by the reams of data, the chairman asks the chief if all of this meant he could, or could not, respond to a major disaster incident.

This chapter has repeatedly emphasized that data are collected in order to be used. Looking at the 3 years of data dealing with fire prevention and inspection taken from three annual reports on the following pages, it should be clear that the department is not using the information. It is impossible to identify trends in most activities. As a manager, can you answer the following questions about the fire prevention and inspection activities for this fire department?

- Is there an active code enforcement program?
- Are violations corrected?

- Are developers installing automatic fire suppression and detection equipment according to code? Do any systems exceed code?
- Who was present at the "fire prevention talks"? Were the presentations appropriate for the audience?
- What type of complaints have been filed? Is there a trend? Were there violations?
- Is the information gathered in inspections incorporated into pre-incident planning?
- Is the fire inspection program effective?
- Is the fire prevention program worth the money spent on materials?
- Are schoolchildren learning fire safety?

MEMO

From: Captain Harry Horton
RE: Fire Prevention & Inspection Report for 1990

	Year-to-Date
Business	30
Educational	12
High Hazards	1
Industrial	15
Institutional	2
Mercantile	20
Place of Assembly	8
Residential	15
Service Stations	8
Occupancies (New)	40
Suppression Systems & Alarms	30
Reinspections	25
Total Inspections	206
Plan Reviews	24
Open Burning Complaints	15
Checkouts (Odors, Smoke)	20
Station Tours	19
Fire Prevention Talks	15
Number of People Attending	1356
Man Hours:	
Inspections	225
Tours	38

Fire Prevention Talks	110
Complaints	35
Misc. Fire Prevention & Inspections	100
Total Man Hours	508

MEMO

From: Captain Harry Horton
Re: Division of Fire
1991 Annual Report
Fire Prevention Report

In 1991 There Were The Following Inspections:

110	Existing Buildings
37	New Occupancy
38	Suppression And Alarm System
26	Reinspections
211	Total

A total of 210 hours were spent on inspections.
32 complaints were checked for a total of 30 hours.
41 fire prevention talks and tours were given to 1556 people using 170 hours.
22 plan reviews were done.
A total of $1375 was spent on fire prevention material in 1991.

MEMO

From: Captain Harry Horton
Re: 1992 Fire Prevention Report

In 1992 there were one hundred and eleven (111) inspections conducted on existing buildings, twenty-five (25) new occupancy inspections, forty (40) suppression and alarm system inspections, and nineteen (19) reinspections for a total of one hundred ninety-five (195) inspections.
A total of two hundred and ten (210) hours were spent on inspections.
Thirty-two (32) complaints were checked for a total of thirty-two (32) hours.
Thirty (30) fire prevention talks and tours were given to 1,276 people using one hundred thirty-seven hours.
Thirty-one (31) plan reviews were done and recommendations made.
A total of $1625 was spent for fire prevention material in 1992.

This illustrates the need to have consistent data collection, recording, and presentation over time. Too often reports are pretty, but they change every year. There is no understanding of *why* the information is in the report. As a result, the presentation of the data changes, the data collected change, and very little is used or useful. The previous examples are actual excerpts from a fire department's annual reports for the years indicated. They are not atypical. Fire officers need to begin to look at data from the standpoint of use. When the use of the data determines the collection and the format, the kind of information gathered and the best way to present it begins to make sense.

The best method of presenting data will vary according to: (1) what is being presented and (2) who the audience is. "A public presentation on fire safety, for example, should be dramatic and clear. It should not concern itself with details, because the best way to have impact in these situations is to make only two or three points and make them stick. For a presentation to council during a budget session, however, much more detail and less sensationalism will have greater impact. The purpose here is to inform and to explain the situation in detail, not to make a lasting impression."[11]

WORKING WITH THE MEDIA

Fire departments often issue news releases through the local media. The use and presentation of data can greatly enhance these announcements. However, it is necessary to be absolutely certain that the information is clear and correctly understood.

It is important to have a good ongoing working relationship with the media. The incident command system identifies a public information officer (PIO) to be the contact point during an emergency. It is a good idea to designate such an individual for nonemergency situations as well. This could be the fire chief or another well-informed officer.

Having a specified spokesperson will help to guide the media to the appropriate person for accurate information. Otherwise, reporters may simply ask the first person they get on the telephone, expecting them to have all of the information. The identity of the spokesperson must be made known inside as well as outside the department. Having a spokesperson or media contact or information officer does not mean that the members of the department are not allowed to express their opinions or answer reporters' questions. However, these responses are *individual opinions, not department positions.* The department position or response comes from the designated source.

If the report is public domain, the media will certainly want a copy of it. In order to focus the media attention, it is a good idea to prepare either a brief fact sheet or a formal news release covering the main aspects of the study. What is important to the department may not seem important to someone who is not very familiar with the fire service. The fact sheet or news release will highlight areas the department views as important.

Once the information is made available, it is a good idea to be available for questions and to encourage dialogue. Should anything in the report be misunderstood, a conversation with the reporter prior to the release of the story to the public may clarify a point. Failing to talk to the media results in missed opportunities to communicate clearly and present the department's position.

If the information is carefully documented and presented in an honest and positive manner, the media as well as the public, fire department superiors, and members of the department are more likely to accept the findings. This is crucial, as a study is most valuable when it is used to shape the future.

Except for the reporting of information, this chapter discusses the internal gathering and use of data and information. The next chapter deals with one specific use of this information: department assessment. The necessity of having good relations with the media is reinforced in that chapter and should incorporate the presentation guidelines reviewed here.

SUMMARY

The use of data is continuing to increase. However, it is also becoming more selective. As information management replaces computer analysis as the focal point, the data will become more valuable since they are more useful. The use of data should be the driving force behind any data collection and processing activities. Data that are inappropriate or unused are often a waste of time.

Local fire department data collected and aggregated at the national level has return benefits in the form of more stringent safety standards and greater public awareness of fire department activities. Data and information can be one of the most valuable tools available to fire officers. They need to be carefully gathered, analyzed, and presented for maximum effectiveness.

CASE STUDY

A fire chief is asked by the township trustees if the extra-long response times from 4 a.m. to 6 a.m. were caused by lack of personnel and whether a standby assignment at the department by part-time personnel could alleviate this problem. The chief responds by presenting a list of hours worked by each member of the department over the past year. In addition, he provides a chart showing who responded to what percentage of calls at what time of day. The township trustees, frustrated, ask the chief to contact an adjoining community to discuss creating an automatic response agreement for the hours in question.

1. What kind of data would have been useful?
2. How would you have presented the data?

ENDNOTES

1. The term "information systems" in this chapter does not refer strictly to computer hardware, but rather to the entire process. Computer systems will be referenced as either computer systems or data processing systems.
2. *NFIRS (National Fire Incident Reporting System) Handbook,* version IV.1 (National Fire Information Council, sponsored by the U.S. Fire Administration, Federal Emergency Management Agency, 1989), p. v.
3. NFPA 901, *Uniform Coding for Fire Protection,* 1990 ed. (Quincy, MA: National Fire Protection Association), pp. 901–6.
4. John R. Hall, Jr., and Arthur E. Cote, "America's Fire Problem and Fire Protection," in *Fire Protection Handbook,* 17th ed. (Quincy, MA: National Fire Protection Association, 1991), pp. 1–23.
5. Ronny J. Coleman, *Residential Sprinkler Systems: Protecting Life and Property* (Quincy, MA: National Fire Protection Association, 1991), pp. 38–39.

6. NFPA 1500 addresses fire fighter health and safety as it relates to the work environment.
7. Carl. E. Peterson, "Collecting Fire Data" in Fire Protection Handbook, 17th ed. (Quincy, MA: National Fire Protection Association, 1991), pp. 10–14.
8. Richard M. Patterson, Jr., Public Safety Services, Inc., Kent, OH, personal interview, April 1, 1993.
9. Richard M. Patterson, Jr., "How to Make Mistakes in 'Computerization'" in *Police Chief,* June 1989.
10. Nancy K. Grant and David H. Hoover, "Statistics in Fire Administration," in *Fire Engineering,* vol. 143, no. 3, March 1990, pp. 93–97.
11. Ibid., p. 97.

CHAPTER 9

Needs Assessment

This chapter outlines the basic steps to follow when conducting a needs assessment. Following these steps ensures that the information necessary is gathered and analyzed in an appropriate manner. This systematic approach can be applied to a variety of decisions facing fire officers. The more it is used, the more familiar fire officers will become with a deliberate and guided means of gathering, analyzing, and using data to assist in identifying the actions that should be taken by the fire department and its members.

The chapter provides several illustrations of the use of needs assessments, but the technique can be used in a variety of ways and in a variety of settings other than those illustrated.

The second section of the chapter presents relevant aspects of gathering data and information. A number of tools and techniques are suggested to assist fire service officers in collecting data in an appropriate manner.

While this chapter does not provide a comprehensive approach to research methods, it does provide the basic groundwork. fire officers who have an interest in needs assessments or other aspects of data analysis are encouraged to take a formal research and statistics course.

WHAT IS A NEEDS ASSESSMENT?

A needs assessment is an analytical and objective identification of service demands in the community and/or required changes in the organization. A needs assessment can be used by fire officers to identify the specific characteristics and extent of a problem and to identify what areas of service need greater emphasis. Chapter 10, Planning, points out that a needs assessment can be performed on: (1) a community, (2) an organization, (3) management, (4) equipment, or (5) an individual. A needs assessment utilizes objective data and proven analytical techniques to document the existence of an organization's operational or service delivery deficiencies. It identifies "true needs" instead of "wants."

Rossi and Freeman (1982) note that social research can be used to make an accurate assessment of the need for various forms of service and suggest alternative strategies for meeting those needs. Social research is research that is based in society rather than in the laboratory. It requires an understanding of the environment (community) and deals with people rather than things. This type of research indicates whether a problems exists, the severity of the problem, and an estimate of the number and type of people affected by the problem. Although a needs assessment is often very descriptive instead of relying on advanced statistics, it is still based on the scientific method and traditional social science research. A needs assessment does not necessarily try to explain the reasons behind needs or predict future events. Instead, it simply documents the situation as it exists.

A needs assessment requires an objective approach that is not biased by any aspect of the participants in the study. If the individual conducting the needs assessment has a vested interest in any of the potential findings, the study becomes suspect. Thus, it is imperative that the study take an unbiased approach and that those conducting the study are able to document all findings and conclusions.

The needs assessment methodology incorporates the use of statistical data to document changes in service demands. This may verify hunches or perceived changes that members of the department have accrued over time. While these perceptions may be accurate, their degree of accuracy is unknown. In addition, if a fire officer cannot demonstrate the truth or facts supporting a perception, it is extremely difficult to convince public administrators, policy makers, and elected officials that things must change. These individuals have to act on solid information rather than intuition. A needs assessment enables the fire department to gather the information in an objective, analytical format that encourages the decision makers in the local government to agree to take action necessary to meet the needs of the department.

WHEN TO USE A NEEDS ASSESSMENT

Needs assessments can be used by fire department administrators to define problems more accurately, to identify the causes of problems, and to examine relative merits of alternative solutions.

A needs assessment should be conducted when the fire department needs to clarify deficiencies in service to the community, such as EMS or public education. This deficiency is generally one of two types, an internal operational deficiency or an external deficiency in meeting the needs of the community.

Internal Needs Assessment

An internal needs assessment focuses on the operation of the department and can be comprehensive or have a narrow range of inquiry. A comprehensive internal needs assessment considers daily operations, management dimensions, and emergency response operations. Each of these three areas can also be considered independently if the department has concerns centered around one aspect more than another. Given the emphasis on emergency response training and preparedness in the fire service, the authors have discovered that most fire departments perform very well in the emergency response mode. It is the daily operations and management dimensions that need improvement. This does not mean that emergency preparedness and response cannot be improved, just that the need for attention is often greater in the nonemergency areas of the fire service.

Management needs assessments focus on the various tasks associated with running the fire department. Management requires skills in a variety of areas and encompasses aspects of the organization such as communication and structure, including, but not limited to, chain of command. A comprehensive management assessment examines many dimensions. However, the number of dimensions is often reduced as a result of preliminary interviews that identify the dimensions that seem to be causing difficulty. The needs assessment is then tailored to fit the specific problems facing the fire department in question. Obviously, it is important that this determination be made by an independent person. Almost everyone has a personal opinion as to what is wrong and everyone has a personal bias that influences the perception of the situation. Thus, the determination of the dimensions to be examined, at least, needs to involve someone from outside the department who does not have a vested interest in the dimensions identified. The outside evaluation might be an independent consultant, a university professor, a citizens' task force, a council committee, or a representative from a regional or state professional association.

Daily operations often include some aspects of management. However, they also involve considering (1) how decisions are made, (2) what assignments are given, (3) how duties are carried out, (4) how fire fighters spend their on-duty time, (5) what maintenance procedures are followed, (6) training opportunities and requirements available, (7) how records are kept, (8) where information is located, (9) in what form information is available, etc. In fact, needs assessment of daily operations centers mainly on information management and personnel on-duty assignments and activities.

A needs assessment of emergency operations incorporates more that just on-scene activities, although that is a major component. This assessment also considers preparation for emergency operations: planning, training, and notification systems. This assessment does not, however, deal with management issues outside fireground operations or emergency medical scenes. A needs assessment is a detailed process designed to identify the root of the problem. (See the discussion of root and symptomatic problems in Chapter 10, Planning.) As such, failure to perform adequately can be a function of miscommunication, inadequate or improper training, lack of planning, refusal to follow commands, panic, or a combination of any of the above. To identify the true root of failure to perform correctly on the fireground, it is necessary to go beyond immediate actions and look at the question of adequate preparedness.

Thus, to some extent the three primary focuses of internal needs assessments—daily operations, management dimensions, and emergency response operations—incorporate aspects of the others. This is because the fire department is a comprehensive unit, and what happens on the fireground affects daily operations and management and vice versa. It is possible, however, to focus the study more narrowly in order to deal with the most serious operational problems. Needs assessments have proven to be a useful tool for improving overall fire department operations and can be used as a foundation for department planning aimed at correcting identified deficiencies. (See Chapter 10, Planning, for further discussion.)

External Community Service Needs Assessment

One of the most prevalent uses of a needs assessment is to identify the types of services and programs the public sector should provide its citizens. In order to do this, it is important to correctly identify the most urgent needs of the community. After all, the fire department exists to provide services to the community on the basis of what it needs, not on the basis of what the fire department or its members want to provide.

These needs assessments are often coupled with program evaluations since needs assessments identify the problem that requires attention and program evaluation checks to see how well the program designed has, in fact, met the need identified. Much of the basic research methodology utilized is the same for a needs assessment and a program evaluation. For example, a survey to determine need must be appropriately constructed and distributed to a random selection of the population. A survey to evaluate the success of a program must meet these same criteria. In addition, once problems are identified, programs are often designed in order to address them. The subsequent evaluation of the program is a modification of the needs assessment, as it determines whether the need still exists or has been mitigated by the program implemented.

An external needs assessment can also be either comprehensive or focused. A comprehensive needs assessment would consider all of the services provided by the department and those desired by the community that would conceivably fall within the responsibilities of the fire department. A more focused needs assessment might consider only emergency medical service to the community. Should it be at basic life support (BLS) or advanced life support (ALS) level? Should the department transport? Should it be a private third service? Another more narrow assessment might consider public education programs and fire prevention activities. In a heavily industrial community, a needs assessment focused on industrial fire protection, OSHA fire brigade regulations, and fire prevention would be appropriate.

Once the focus of the external needs assessment is identified, it proceeds along the same steps as the internal needs assessment. It is important to identify exactly what the focus of the assessment is before inception. Otherwise, one of two things is likely to occur: (1) more questions will come up during the study and it will have to be expanded, or (2) the findings will not address the concerns for the appropriate service that prompted the study in the first place. Identifying what type of needs assessment to conduct and the specific focus of the activity is crucial to the success of the endeavor.

Once the parameters of the needs assessment are set, the actual study process can begin. It is important to follow the requisite steps in order to maximize the accuracy and benefit of the process. Appropriate research methodology must be utilized in gathering the data if the data are to be valid and provide a true picture of the situation at hand. This will be addressed in more detail after the steps of the process have been presented.

HOW TO CONDUCT A NEEDS ASSESSMENT

There are a number of specific steps inherent to any needs assessment, regardless of the type or focus. These will be discussed in detail in this section of the chapter and reviewed in Chapter 10, Planning, since a needs assessment is often conducted as an integral part of the comprehensive planning process. The orientation of the persons conducting the study, the members of the organization, and the community involved are as important as the specific steps in the process.

The Importance of a Positive Approach

The fire department exists to provide a needed service to the community in the most efficient and effective manner possible. A needs assessment is an analytical research tool used to improve the efficiency and effectiveness of department operations. Its purpose is to pinpoint areas that need attention to elevate the quality of service delivered—to identify areas in department operations that need additional resources, restructuring, etc. In this way the delivery of services can better match the needs of the community. In addition, internal needs assessments can enhance the efficiency of service by improving internal operations.

Remember that a needs assessment is a tool to help improve the fire department. It should never be conducted for the purpose of castigating or punishing. The use of needs assessments demonstrates interest in improving services and maximizing operations by identifying and correcting difficulties and building on success.

Steps in Conducting a Needs Assessment

The specific steps in conducting a needs assessment and a brief explanation of each step are listed below. Following this listing are sample needs assessments, which provide detailed information on each step.

1. *Define the focus of the assessment.* This sets the parameters of the study. Example: Is it the entire organization or a unit within the organization?
2. *Collect data.* Gather *all* existing data and information on the subject of the study. This includes biased as well as objective data and individual opinion as well as documented fact. (These may be indicators of root versus symptomatic problems.) If biased information is collected, however, it must be identified, analyzed, and interpreted as such.
3. *Generate data.* If data or information are not available, it may be necessary to collect primary data through research techniques such as surveys and focused interviews. (**Secondary data** are data someone else has collected for a different purpose, but which one can use in the study. **Primary data** are data gathered specifically for the study currently taking place.)
4. *Organize data.* The data must be organized in logical groupings and format so that it can be constructively utilized. This organization of data must be done with the focus of the study

in mind. The data are grouped in ways that assist in providing answers to the questions of the study.

5. *Analyze information—compare and contrast data.* Data analysis can be very sophisticated; however, that is not necessary in order to have a valid study. What is important is that the data be analyzed *appropriately.* To compare and contrast data is a fundamental analytical technique. Basically, this is a system of identifying similarities and differences, of identifying trends or patterns in information.
6. *Interpret data.* Once the data have been analyzed, the results must be translated into a form useful to the organization. It must relate to the needs of the subject of the study.
7. *Determine needs.* Once the data have been interpreted and the problems identified, it is necessary to plan what needs to be done to rectify the problems. The specific needs have to be defined in a form that identifies department services or operations that could provide solutions.
8. *Set priorities.* The final step in the needs assessment is to assign a priority of needs on the basis of which are most critical to the provision of quality service. Once the needs are given priority ranking, the department can begin to plan activities.

This systematic approach to identifying problems can help a department properly identify service needs. In addition, the structured approach results in the accumulation and analysis of data in a manner enabling the department to demonstrate the reasons for its conclusions and to defend its decisions.

The following examples illustrate how a needs assessment procedure could be used in different settings.

Needs Assessment #1: Increase in Number of Fires

At the end of August, as he is preparing his monthly fire activity report, Chief Jeffrey confirms what many fire department members have been saying, that there was a large (25 percent) increase in the number of fire calls between mid-June and August. This fact was reported during the monthly meeting. A discussion of the situation ensued, during which a number of members offered different reasons for the increase and different solutions. Instead of jumping to conclusions, however, Chief Jeffrey appoints a study committee composed of members of the department to undertake a needs assessment on the problem. They will have 1 month to perform the study and then will report back to the members of the fire department at the next monthly meeting. The committee takes the following steps:

1. Define the Focus of the Study. The major question of the study is: Was there an increase in the number of fires during the summer this year? What type of fires were they? What caused the fires? If there was an increase, how can the fire department reduce the number next year?

2. Collect Data. The department keeps a run sheet as well as a completed NFIRS (National fire Incident Reporting System) form for each incident. The chief summarizes the numbers in his monthly reports, but does not count the fires by type. The committee decides to review the last 2 years and the current year for the months of January through August and collect the following data from the run sheets and NFIRS reports:

- Type of fire
- Location of fire
- Cause of fire
- Amount of property damage

During this process, they discover that the cause had not been recorded for all of the incidents this year. Another member of the fire study committee, a local farmer, believes the drought had a lot to do with the number of fires and decides to collect weather data for the last 3 years. Specifically, he collects:

- Number of days of rain in May, June, July, and August of each year
- Total rainfall for the year as of May 31, June 30, July 31, and August 31 of each year
- Average temperature for May, June, July, and August for each year
- Number of days with temperature above 90°F for June, July, and August of each year

3. Generate Data. Because the existing data showed a 25 percent increase in the number of fires over the previous year and a 15.8 percent increase over the year before that, the committee decides to identify the cause of those fires listed as unknown that took place this year. Therefore, the committee decides to interview owners of the property on which the incident occurred and the person who first reported the incident.

Because they do not know what to expect, the committee members conduct an open-ended interview with only three main questions:

(a) Do you remember the fire on _________ (date) that occurred at _________ (location)? [This is asked to refresh the interviewee's memory and identify exactly what is being asked.]
(b) Did you notice anything unusual before the fire started?
(c) Do you know what caused the fire/materials to ignite?

The committee knows the last question is simply one of perception, but believes it is important to identify supposed causes since there is no way to investigate the scene at this date.

TABLE 9-1 Type of Fires—1990, 1991, 1992

NFIRS Number	*Type of Situation Found*	*1990*	*1991*	*1992*
11	Structure fire	7	11	9
13	Vehicle fire	4	5	7
14	Trees, brush, grass fire	8	9	19
35	Extrication	5	4	2
41	Spill, leak with no ignition	3	2	2
45	Arcing, shorted electrical equipment	2	1	3
61	Smoke scare	3	4	4
73	System malfunction (alarm drop)	6	8	9
	TOTAL	38	44	55

4. Organize Data. The committee then organizes the data in a comparative format. They put together a number of charts and tables, such as Tables 9-1 to 9-3 and Figures 9-1 and 9-2.

TABLE 9-2 Cause of Tree, Brush, Grass Fires—1992

NFIRS Number	*Ignition Factor*	*Number of Incidents (Total = 19)*
00	Ignition factor undetermined or not reported	10
21	Suspicious, not during civil disturbance	2
31	Abandoned, discarded material	2
49	Misuse of material ignited—fireworks	3
84	Lightning	2

FIGURE 9-1 Division of Fire

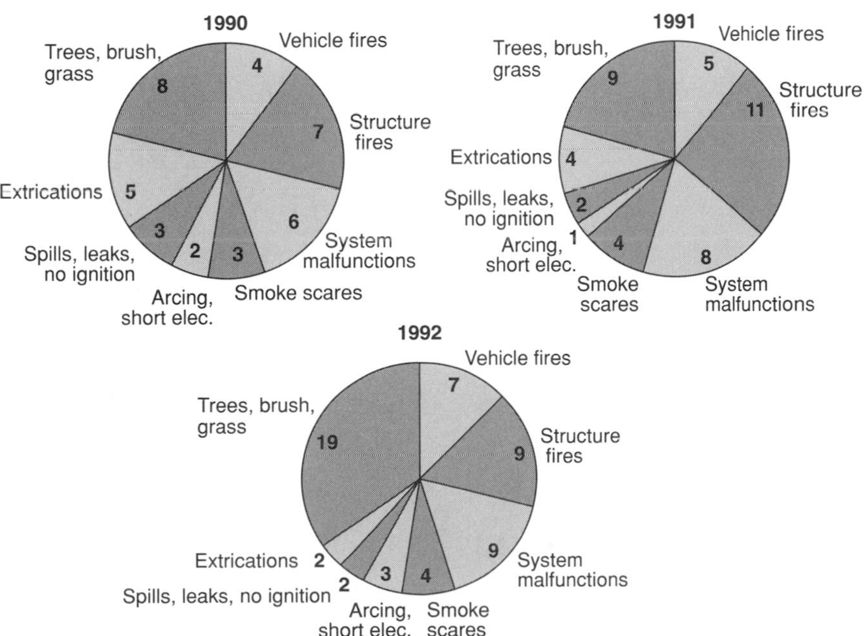

TABLE 9-3 Interview Information for Fires of Undetermined Origin

Incident Number	*Question 1: Remember fire*	*Question 2: Noticed unusual*	*Question 3: Suspected cause of fire*
127	Yes	Loud bangs—guns or fireworks	Don't know—could be fireworks
136	Yes	Family picnicking—no campfire	Don't know
149	Yes—owner Yes—reported	No—not home Kids lighting fireworks	No idea Kids playing with fireworks
161	Yes	No	Don't know
162	Yes	No	No
171	Yes—owner Yes—child reported	No—lots of noise Admitted setting off fireworks	No idea Fireworks
177	Yes	Lightning—no rain	Lightning
190	Yes	No—it started an hour after the fireworks display	No idea, supposed match or fireworks could have smoldered for an hour
195	Yes	No—many people smoking in the area	Cigarette or match
202	Yes	Sharp, loud noises; could have been fireworks	Don't know; could have been fireworks

FIGURE 9-2 Causes of Fires Compared, with and without Information from Interviews

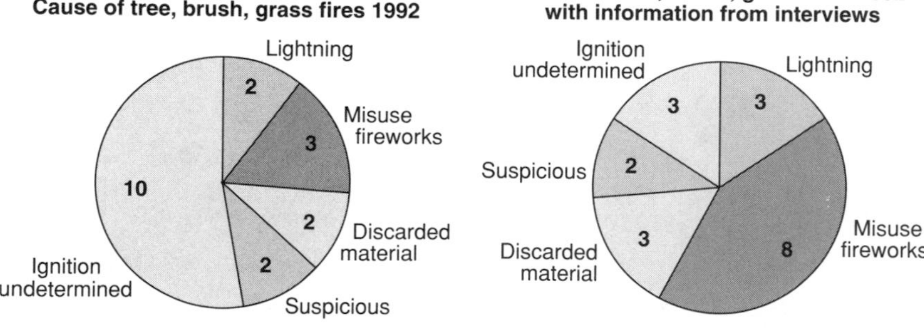

5. Analyze Information—Compare and Contrast Data. The data demonstrate a definite increase in grass fires. The primary cause of grass fires in the original run data was identified as "ignition factor undetermined." The interviews with individuals concerning the fires caused by ignition factors undetermined indicate that some of these were actually the result of discharging fireworks. The number of grass fires increased tremendously in the last year. The area is definitely experiencing drought conditions.

Table 9-3, Fires of Undetermined Cause Interview Information, demonstrates that 50 percent of the fires (five out of ten) were probably due to fireworks. Only three remained unknown because two others were given probable causes: lightning and cigarettes. Thus, with this five and the other three fires originally identified as caused by fireworks, the data demonstrate that eight fires (42 percent of all grass fires this year) were caused by discharge of fireworks.

6. Interpret Data. There has been an increase in the number of fires caused by children using fireworks in the community this year. This has resulted in an increase in fires. The dry conditions of the area facilitate ignition.

7. Determine Need. The fire department and the community need to address the issue of unsupervised use of fireworks. The fire department can begin to address this by including a section in its public education programs in the school. In addition, the chief can make the township trustees aware of the situation. The trustees can then decide whether they want to pursue the establishment of an ordinance to deal with the issue.

The fire department can do nothing to alleviate the dry conditions caused by the weather. However, they can try to make the public more aware of the ease with which fires can be ignited in drought conditions, hopefully improving safety practices within the community.

8. Prioritize Needs. Given the potential for injury associated with the use of fireworks, the fire department decides to concentrate their public education campaign in the schools. Since school has just started, they will incorporate a section on the dangers of fireworks in their annual presentation and put together an information flyer for students and their parents.

Also, the fire chief and a member of the fire department study committee will meet with the township trustees to present their findings. (Although the members of the department might want to advise the township to enact an ordinance dealing with the sale and discharge of fireworks, that action must be taken by the officials elected to make the policy in that area. Therefore, in this case the township trustees will have to be persuaded to take action.)

This case study illustrates how one fire department followed the steps of a needs assessment in order to solve a very real problem. The methodology does not have to be sophisticated, but it does need to be orderly and accurate. Keeping the steps in mind helps to order the process and prevent missing potentially useful information. When gathering data and information, however, there are a number of things to remember.

Sometimes, once part of the question is answered in the information-gathering process, further questions arise. In this case, it is necessary to repeat some of the data-gathering and analysis steps for the new set of questions. This situation is illustrated in the next scenario.

Needs Assessment #2: Cultural Differences

Recently, numerous immigrant families have moved into a central city. The fire department is experiencing increased fire calls in these areas and argues that these additional calls demonstrate a need for more personnel. Unfortunately, the safety director informs the department that there is no money available.

During the coffee break at the monthly officer's meeting, some of the officers discuss several unusual runs they've experienced recently. They note that some of the fires were related to the practice of using charcoal or gas grills to cook in the living room. Also, several families have changed the wiring in their homes and are overloading their systems. fire Prevention Captain Herzog overhears part of this conversation and realizes that something must be done.

In discussing this with the assistant chief in charge of EMS, Captain Herzog discovers another trend of growing concern among EMS officers. According to the paramedics, a high proportion of the calls made to the newer immigrant homes require Advanced Life Support (ALS) care.

The captain and assistant chief decide to approach Chief Sanchez with their discoveries. The first thing the chief asks is, "Are you sure this is a real trend or just a few fire fighters talking? We are spread so thin that I am going to make sure we don't jump into action on a few hearsay conversations."

1. Define the Focus of the Needs Assessment. This needs assessment is designed to determine whether there is an increase in calls to areas where immigrants live. It has a secondary purpose of (1) determining the cause and origin of fires and (2) determining whether a higher proportion of EMS calls are severe in these districts. It will then be necessary to determine why.

Questions to be answered:

- I. Has there been an increase in the number of calls to immigrant neighborhoods from the previous 2 years?
  - A. Is this increase greater than that in the other neighborhoods in the city? (Or is it that there is an increase in calls everywhere but fire fighters just notice these calls more?)
- II. How many of the fires in these districts were related to using charcoal or gas grills for indoor cooking? How many of the fires were related to electrical wiring overloads?
- III. What proportion of calls to immigrant families require ALS care? What proportion were transported to a trauma center?
  - A. How does this proportion compare with calls to other segments of the population?
  - B. Is age a factor in the calls? (Are more of the victims elderly?)

Captain Herzog agrees to research answers to Questions I and II while the EMS assistant chief investigates Question III. (These will be presented separately so that the steps are obvious and the procedures of the needs assessment can be clearly identified.)

For Question I:

2. Collect Data. For Question I, Captain Herzog asks the lieutenant in charge of records to pull a listing of the number of calls by neighborhood over the past 3 years. The city is divided into 13 districts, and each run is logged according to the district in which it was located. In order to verify the districts in which the majority of immigrants are located, Captain Herzog meets with the director of social services and a member of the city planning department. They all agree that most of the immigrant groups live in Districts 5, 8, and 13.

3. Generate Data. No data need to be generated. All needed data are currently collected for other uses.

4. Organize Data. The information from the run records is put into Table 9-4.

TABLE 9-4 Fire Calls by District for Current and Previous 2 Years

District	*1991 number*	*1992 number*	*Percent change*	*1993 number*	*Percent change*
1	56	38	−32.1	39	+2.6
2	72	68	−5.5	73	+7.4
3	78	79	+1.3	77	−2.3
4	65	63	−3.1	67	+6.3
5*	65	77	+18.5	88	+14.3
6	70	69	−1.4	68	−1.4
7	72	69	−4.2	73	+5.8
8*	75	82	+9.3	87	+6.1
9	70	64	−8.6	65	+1.66
10	20	21	+5.0	20	−4.8
11	44	40	−9.0	37	−7.5
12	70	68	−2.8	64	−5.9
13*	60	70	+16.7	80	+14.3
Total	817	808	−1.1	838	+3.7

*Immigrant district

5. Analyze Information—Compare and Contrast Data. The data show a slight overall increase in the number of fires for the current year. The department has had more fires at this point in time than it did over the last 2 years. The last year, however, saw a minor decrease in the total number of fires.

The proportion of the increase is much higher in the districts in question than in other districts of the city. These three districts show a steady increase in fires over the past 3 years.

6. Interpret Data. There is a definite increase in the number of fires in the neighborhoods being studied. These fires may be related to cultural practices; however, these are also some of the older neighborhoods in the city, and older areas tend to have more fires. Since the records do not identify the race or ethnicity of the victims of a fire incident, it is not possible to say with certainty that the fires are taking place in minority residences. There does seem to be a reasonable demonstration of need to find out why these fires are taking place and to try to do something about it, however.

7. Determine Needs. Given the information at hand, the need is to go to the next step and find out why there is an increase in fires in order to identify any patterns that may be taking place. Once that additional information is gathered, it may be possible to identify the need more specifically. At this point all the researcher can conclude from the data is that there is a need for more fire prevention directed toward these neighborhood, but what type is uncertain.

8. Prioritize Needs. Until the problem is identified in more detail, no other need can be addressed.

For Question II:

1. Define the Focus of the Needs Assessment. How many of the fires in these districts were related to using charcoal or gas grills for indoor cooking? How many of the fires were related to electrical wiring overloads?

2. Collect Data. Captain Herzog asks the lieutenant in charge of records to identify the cause and origin of all fires in Districts 5, 8, and 13 over the past 3 years. Because neither the race nor the ethnicity of the victims is entered on the run sheets, the lieutenant cannot identify whether the incidents in question took place in immigrant residences.

3. Generate Data. Since neither the race nor the ethnicity of the residents can be gathered from the run sheets, it might be helpful to verify that the incidents did take place in immigrant residences. However, this would be a labor-intensive activity and the staffing is not available. Instead, Captain Herzog decides to take a different approach, depending on the findings.

4. Organize Data. The data on the cause and origin of the fires are placed in tables and charts to illustrate the findings. (See Tables 9-5 to 9-7 and Figures 9-3 and 9-4.)

TABLE 9-5 Cause of Fires

	1991	*1992*	*1993*
Electrical	16% (131)	18% (146)	20% (168)
Open flame (candle, fireplace, etc.)	10% (82)	9% (72)	11% (92)
Smoking	14% (114)	12% (98)	10% (84)
Cooking related	12% (98)	14% (113)	15% (126)
Children playing with matches	9% (74)	8% (65)	9% (76)
Fuel ignition (gas, propane, kerosene)	11% (89)	10% (80)	12% (100)
Natural causes (e.g., lightning, sun)	11% (89)	15% (121)	12% (100)
All others	17% (140)	14% (113)	11% (92)
Total	100% (817)	100% (808)	100% (838)

NOTE: Percentages may not equal total due to rounding.

FIGURE 9-3 Comparative Percentage of Cooking Fires

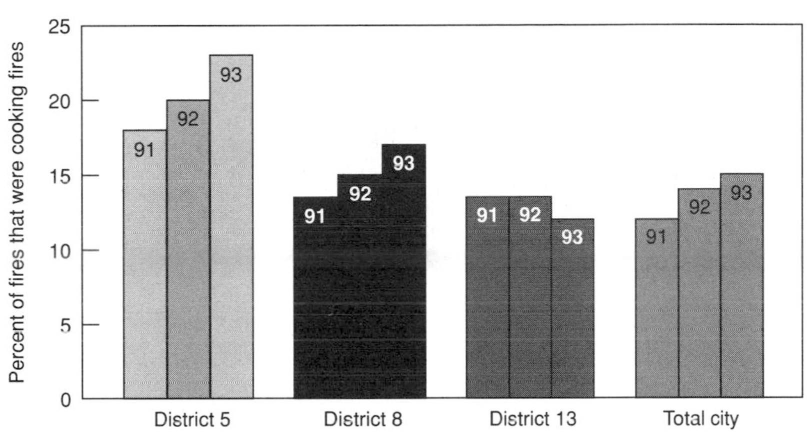

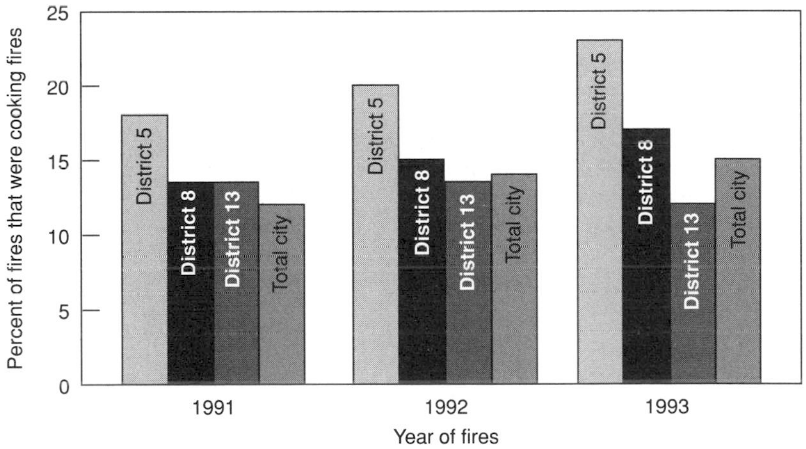

FIGURE 9-4 Comparative Percentage of Electrical Fires

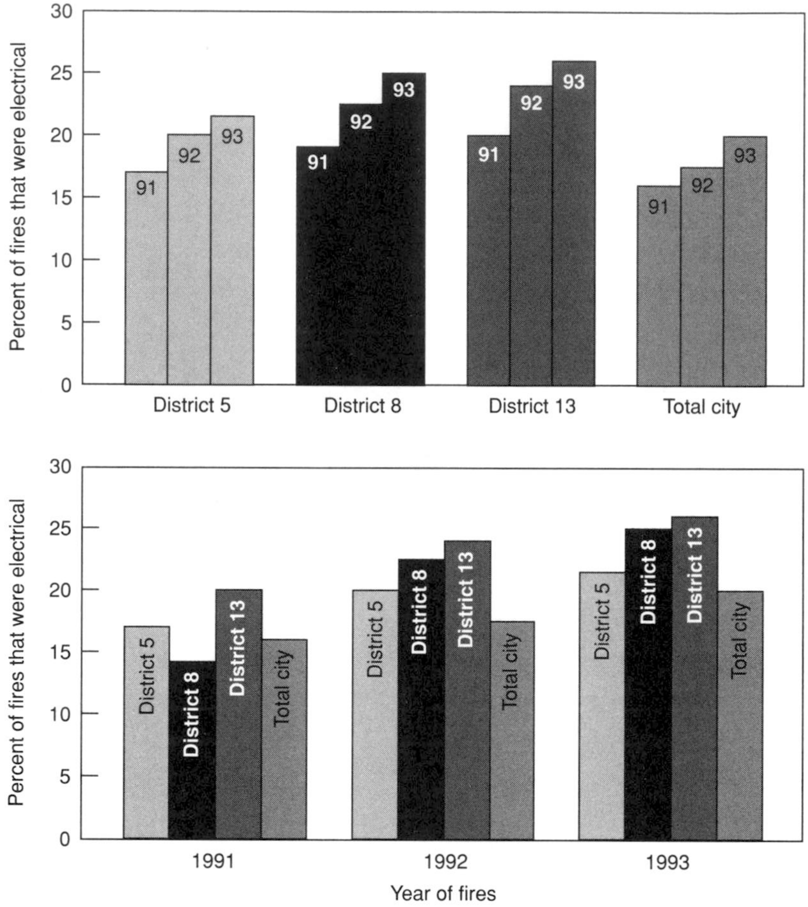

TABLE 9-6 Cause of Fires in Districts 5, 8, and 13

	1991	*1992*	*1993*
District 5			
Electrical	17% (11)	19% (15)	20% (18)
Open flame (candle, fireplace, etc.)	11% (7)	9% (7)	11% (10)
Smoking	14% (9)	13% (10)	11% (10)
Cooking related	18% (12)	19% (15)	22% (19)
Children playing with matches	5% (3)	6% (5)	7% (6)

TABLE 9-6 (continued)

	1991	*1992*	*1993*
Fuel ignition (gas, propane, kerosene)	14% (9)	12% (9)	13% (11)
Natural causes (e.g., lightning, sun)	11% (7)	12% (9)	10% (9)
All others	14% (7)	9% (7)	6% (5)
Total	100% (65)	100% (77)	100% (88)
District 8			
Electrical	19% (14)	22% (18)	24% (21)
Open flame (candle, fireplace, etc.)	8% (6)	9% (7)	9% (8)
Smoking	15% (11)	15% (12)	13% (11)
Cooking related	13% (10)	15% (12)	17% (15)
Children playing with matches	8% (6)	10% (8)	8% (7)
Fuel ignition (gas, propane, kerosene)	9% (7)	11% (9)	9% (8)
Natural causes (e.g., lightning, sun)	8% (6)	10% (8)	10% (9)
All others	20% (15)	10% (8)	9% (8)
Total	100% (75)	100% (82)	100% (87)
District 13			
Electrical	20% (12)	24% (17)	25% (20)
Open flame (candle, fireplace, etc.)	10% (6)	10% (7)	11% (9)
Smoking	13% (8)	13% (9)	13% (10)
Cooking related	13% (8)	13% (9)	13% (10)
Children playing with matches	8% (5)	10% (7)	11% (9)
Fuel ignition (gas, propane, kerosene)	12% (7)	10% (7)	11% (9)
Natural causes (e.g., lightning, sun)	10% (6)	10% (7)	8% (6)
All others	13% (8)	10% (7)	9% (7)
Total	100% (60)	100% (70)	100% (80)

NOTE: Percentages may not equal total due to rounding.

TABLE 9-7 Comparative Percentage of Cooking and Electrical Fires

	1991	*1992*	*1993*
Cooking fires			
District 5	18% (12)	20% (15)	22% (19)
District 8	13% (10)	15% (12)	17% (15)
District 13	13% (8)	13% (9)	12% (10)
Total City	12% (98)	14% (113)	15% (126)
Electrical fires			
District 5	17% (11)	20% (15)	21% (18)
District 8	19% (14)	22% (18)	25% (21)
District 13	20% (12)	24% (17)	26% (20)
Total City	16% (131)	18% (146)	20% (168)

NOTE: Percentages may not equal total due to rounding.

5. Analyze Information—Compare and Contrast Data. The distribution on cause and origin of the three immigrant districts in question versus the distribution of the city as a whole demonstrates a higher proportion of cooking and electrical fires in the neighborhoods under study. There is some support for the claims of the fire fighters that people in Districts 5, 8, and 13 are not following safe practices when cooking and that some may be adapting wiring or overloading electrical circuits.

Because reports on the cooking fires do not specify the type of cooking appliance, Captain Herzog asks the lieutenant to produce a cross-tabulation of fires caused by cooking accidents sorted by the location of the incident in the house. (This is a return to Step 2, Generate data.) He then creates a table showing this information (Step 4, Organize data). (See Table 9-8.)

On the basis of this cross-tabulation, there is some evidence that families in District 5 have cooking accidents in the living room as well as the kitchen.

6. Interpret Data. A higher-than-expected number of fire incidents in immigrant neighborhoods are caused by cooking accidents. The records do not show that these accidents were specifically caused

TABLE 9-8 Location (Origin) of Cooking Accidents for 1991, 1992, 1993, Districts 5, 8, and 13

1991	*Location*	*District 5*	*District 8*	*District 13*
	Kitchen	58% (7)	80% (8)	88% (7)
	Patio	17% (2)	10% (1)	12% (1)
	Living room	17% (2)	0% (0)	0% (0)
	Other	8% (1)	10% (1)	0% (0)
		100% (12)	100% (10)	100% (8)
1992	*Location*	*District 5*	*District 8*	*District 13*
	Kitchen	60% (9)	83% (10)	78% (7)
	Patio	7% (1)	8% (1)	11% (1)
	Living room	33% (5)	0% (0)	11% (1)
	Other	0% (0)	8% (1)	0% (0)
		100% (15)	100% (12)	100% (9)
1993	*Location*	*District 5*	*District 8*	*District 13*
	Kitchen	47% (9)	87% (13)	80% (8)
	Patio	11% (2)	7% (1)	10% (1)
	Living room	37% (7)	0% (0)	10% (1)
	Other	5% (1)	7% (1)	0% (0)
		100% (19)	100% (15)	100% (10)

NOTE: Percentages may not equal total due to rounding.

by cooking on charcoal or gas grills in the living rooms, but there is some evidence that this may be true, i.e., the number of cooking fires that originated in the living room and dining room as opposed to the kitchen.

There are an unusually large number of electrical fires in Districts 8 and 13. Some of these are identified as being caused by wiring problems, overloading circuits, and attempts to steal electric energy from other sources. Many of these electrical fires are not attributed to any specific cause, however.

In order to verify that a substantial number of the homes involved in fire incidents are occupied by immigrants, Captain Herzog decides to meet with the directors of social service agencies that serve the people in Districts 5, 8, and 13. This includes the directors of the public health centers located in the areas. In addition, Herzog meets with the priests of some of the immigrant religious organizations in the city. During these meetings he asks about the problems of fire in the community and about the cultural customs of the residents that could pose a fire hazard. (Repeat of Step 2.)

7. Determine Needs. There is a definite need to address fire prevention efforts in the three districts. There needs to be an increased awareness among the members of these communities concerning the dangers of improper cooking practices and of inadequate wiring or misuse of electrical supply devices. The information gathered from the cultural experts during the interview convinces Captain Herzog that whatever type of program the fire department wants to pursue, they need to incorporate some cultural sensitivity to get the members of the community to accept the safety messages.

8. Prioritize Needs. The fire department needs to work with the immigrant community to determine how best to present a safety program. Once this is done, public awareness concerning the dangers of cooking and inappropriate use of electricity and inadequate wiring must be increased. Who should do this or how is something that needs to be planned. At this point, however, the need is clearly identified.

For Question III:

1. Define the Focus of the Needs Assessment. What proportion of calls to immigrant families require ALS care? What proportion were transported to a trauma center?

A. How does this proportion compare with calls to other segments of the population?
B. Is age a factor in the calls? (Are more of the immigrant victims elderly?)

2. Collect Data. The EMS runs are collected on a different form and include more information concerning patient characteristics. Therefore, the lieutenant is asked to select the EMS responses made to immigrant patients and to compare the care given to these patients with that given to other patients. He is also asked to divide the response by age in order to look at the number over 65 and under 65. Since the assistant chief is just interested in whether more of these calls are ALS, not whether there has been an increase, he is only interested in looking at the calls for that year.

3. Generate Data. No data need to be generated at this point.

4. Organize Data. The data are organized into tables. (See Table 9-9.)

5. Analyze Information—Compare and Contrast Data. The nature of care given to the immigrant patients is compared with that given to other patients in the city. The data clearly demonstrate that a higher *proportion* of calls to these patients require ALS care. In addition, when the age factor is controlled, it is clear that even younger immigrant patients required advanced care.

6. Interpret Data. For some reason, EMS calls from immigrants had a higher probability of requiring advanced care. This is true for younger as well as older patients. From this data, it is impossible to say why.

TABLE 9-9 Comparative Level of Treatment on EMS Calls

	Immigrant Group #1 (10%)	*Immigrant Group #2 (8%)*	*All Others (82%)*
Under 65—71% of Patients (2320)			
ALS	32% (74)	28% (52)	10% (190)
BLS	68% (158)	72% (134)	90% (1712)
Total	100% (232)	100% (186)	100% (1902)
Over 65—29% of Patients (948)			
ALS	35% (67)	38% (54)	17% (105)
BLS	65% (123)	62% (88)	83% (511)
Total	100% (190)	100% (142)	100% (616)

7. Determine Needs. The needs developed from this information could be varied. For example:

(a) There is a need to educate immigrants about the proper use of the EMS system.
(b) There is a need to introduce more preventive care practices among Asian-Americans.
(c) There is a need to inform the leaders of the city and of the immigrant communities about the high proportion of severe accident and illness calls.
(d) There is a need to formally notify fire fighter/paramedics of the probability of requiring ALS when responding to patients in Districts 5, 8, and 13.

8. Prioritize Needs. Given the mission of the fire department and the lack of staffing, the needs *for the fire department* would probably be prioritized as follows:

(a) There is a need to formally notify fire fighter/paramedics of the probability of requiring ALS when responding to the identified patients.
(b) There is a need to inform the leaders of the city and of the immigrant communities about the high proportion of severe accident and illness calls.
(c) There is a need for education about the proper use of the EMS system.
(d) There is a need to introduce more preventive care practices.

The fire department would probably address the first and second needs and relegate the others to a later point in time or simply let other members of the city and community become aware of the needs and see if they are willing to use their resources to address those needs.

This case study demonstrates how needs assessment can address more than one question at a time. As long as the subjects of the questions are related, a variety of data can be gathered and analyzed to assist in solving the problem. Being aware of the community and the department is also important. For example, knowing about the different cultural communities enables fire officers to target an investigation. Interacting with other public administrators facilitates access to information and can be used to help verify suspicions. And understanding the capabilities of the department's information management system lets one know exactly what to expect from data on hand in the department. It is important to stay focused and answer the question of the study without getting sidetracked along the way. Once the needs are identified, a plan can be enacted to address them.

Needs Assessment #3: Where to Start Public Education

The combination fire department receives a substantial gift from a prominent family in memory of their grandchild, who was killed in a fire caused by children playing with matches. The money will support the addition of a public education specialist to the contingent of full-time fire fighters and also cover the cost of public education materials.

The gift specifies that the money and programs must first be delivered to areas where it will "do the most good in saving the lives of innocent children." In order to accurately identify this area—and document the decision in a manner that will be acceptable to the donors—Chief Marks decides that the department should identify the area of greatest risk and make that the main focus of public education programs. The chief contacts a local university, which provides a student intern to conduct the study.

1. Define the Focus of the Needs Assessment. The student decides that there are two major questions to be answered: (1) What areas of town have the highest incidence of children playing with matches? (2) What are the ages of the children involved in the incidents? The answers to these two questions can be used to identify which schools and preschools or day care centers should receive the first programs. Depending on the ages of most of the children, the public education specialist with the most experience with that age group would be best qualified.

2. Collect Data. The student intern gathers the fire reports from the preceding 2 years.

3. Generate Data. No additional data need to be generated.

4. Organize Data. The student identifies the attendance zones for the public schools in the city and organizes the incidents of children playing with matches by the schools. She also classifies the incidents according to the ages of the children involved. She then identifies the ages within each school zone. (See Table 9-10.)

TABLE 9-10 Incidents of Children Playing with Matches in the Past 2 Years

School Attendance Zone	*Total Incidents*	*0–2 Years*	*3–4 Years*	*4–5 Years*	*6–7 Years*	*8–9 Years*	*10–12 Years*
Washington School	22	2	10	4	2	2	2
Jefferson School	18	1	3	4	8	2	0
Roosevelt School	8	0	1	3	3	0	1
Adams School	6	0	3	0	2	1	0
Kennedy School	9	1	0	4	3	0	1

5. Analyze Information—Compare and Contrast Data. The student discovers two areas with the highest number of incidents. In the Jefferson School zone, most children involved in the incidents are 6 to 8 years old. In the Washington School attendance zone, however, the children involved are considerably younger, 3 to 5 years of age. There are incidents of children playing with matches in all of the school attendance zones of the city. The intern ranks these zones according to the number of incidents in each.

6. Interpret Data. There is a need for public education for children in order to teach them fire safety. There is a difference in the rate of incidents according to location in the city.

7. Determine Needs. There is a need for both preschool and elementary school fire safety programs in the city. Specifically, fire safety needs to be taught to preschool through 6th-grade children.

8. Prioritize Needs. The first priority should be the Jefferson and Washington attendance zones. Thus, the student intern recommends that the fire department target the K–3 grades in the Jefferson School. In the Washington school attendance zone, where the majority of the children involved in incidents are preschoolers, the focus will be on preschool education offered to day care programs. Priority for hiring should be given to a public education specialist with experience working with preschool and elementary school students.

Needs assessments are often performed by someone who does not work full-time for the fire department. Sometimes the objective outlook can provide a fresher, more unbiased view of the situation. Needs assessments can be uncomplicated and straightforward or extremely complex. In either case, however, the important aspect is that the procedures are followed correctly.

THE USE OF DATA AND INFORMATION

Personnel involved in needs assessments or evaluations are advised to take a course in research methodology. Basic knowledge of how to gather and use data is very beneficial even if fire officers are not researchers. In addition, understanding various reports can help officers ask the right questions and build a case for or against the information in the report. In order to do this, a few elements of data collection and use should be remembered.

The Importance of Truthful Statistics

Never make up statistics. It is always possible to check and double-check information. Researchers, lobbyists, and marketing experts all rely on statistics. They do not use false statistics because it is unethical and unprofessional. A lobbyist or researcher who uses false statistics and is caught has ruined his or her career. It is also possible to mislead people with accurate statistics. This occurs more frequently than falsifying numbers, and fire officers need to be able to recognize misleading reports.

The way in which statistics are presented can change the way the information is perceived. Even the use of adjectives can make a difference. For example, "Over two-thirds of the voters, 66.1 percent, support the candidate" suggests a much greater level of support than does, "With almost one-third of the voters in opposition, the candidate has support from 66.1 percent of the voters."

In addition, perceptions can be shaped by holding back statistical information rather than sharing it. For example, in one community a group of citizens pushing for a full-time paramedic-level emergency medical service pointed out that 20 percent of the calls in 1 month had a response time of 8 minutes or more. Upon further investigation, it was determined that these calls consisted of two out of ten for that month and were 50 percent (or two of four) of all calls of 8 minutes or more for the last 12 months! (Neither call required advanced life support or transport.) The citizens didn't lie; they simply shared information selectively. Understanding statistics helps fire officers understand arguments, counter a position, and present their own position. As fire service administration becomes more sophisticated, this skill becomes increasingly important.

Always look at statistics and data critically. Understand exactly what is being presented and how the interpretation is being made. Also find out where the statistics come from and how they are gathered. This will indicate whether there is a perception or bias involved in the information.

Confidentiality

When surveying or interviewing individuals, the issue of confidentiality will need to be addressed. In research, *confidentiality* is the assurance that particular information or responses will not in any way or at any time be attributed to or linked with any specific individual who participates in supplying information. This is crucial to gathering honest information and maintaining the ethics associated with conducting research.

When participants in a study realize that superiors or the public may find out who gave certain information or held certain opinions, they are intimidated and may either not share information or give false information. This does not help the fire department define problems and address needs. In instances where the researcher wants to share information from one of the participants and wishes to cite the source of the information, the researcher must gain a release from the participant. This release is a letter of agreement stating that the information identified may be attributed to the interviewee who participated in the study. An example of two letters that may be used follows.

LETTER TO PARTICIPANT IN INTERVIEW

Captain Harold Jones
Haz Mat Officer
Upper Overshoe Fire Department

Dear Harold Jones:

On November 5, 1992, you participated in an interview. This interview was part of a study of haz mat officers throughout the state. We are now in the process of writing up the findings of the study and would like to quote you in the report.

For one of the questions in the study you rated the cooperation between the police department and the haz mat team as high. We asked:

> To what do you attribute this cooperation?

We would like to quote your answer:

> "I am convinced that we work together so well because we trained together. We all took the same awareness course and then we each shared our view of an incident and what we thought our responsibilities were. Then we worked together as part of the training in different drills and exercises. Since we understand what each other does, we don't try to upstage or get in the way."

If you agree to give us permission to use this quote and attribute it to you, using your name and the name of your department, please sign the enclosed letter and return it to us.

We thank you for your cooperation in this matter.

Sincerely,

Dr. Sally Feldin
Research Associate

LETTER FROM PARTICIPANT GIVING PERMISSION

Dr. Sally Feldin
Research Associate
State Office of Emergency Management

Dear Dr. Feldin:

I hereby give you permission to use the quote located at the end of this letter in your report of the study and in any future publications. You may attribute the statement to me and identify me as a member of the Upper Overshoe fire Department.

Quotation that may be used:

> "I am convinced that we work together so well because we trained together. We all took the same awareness course and then we each shared our view of an incident and what we thought our responsibilities were. Then we worked together as part of the training in different drills and exercises. Since we understand what each other does, we don't try to upstage or get in the way."

Name of Individual: Captain Harold Jones
Organization of Individual: Upper Overshoe Fire Department

Signature Certifying Permission: ____________________

Date Permission Granted: ____________________

DATA-GATHERING TECHNIQUES

Among the large variety of data-gathering techniques are two main types: the use of existing data sources and the gathering of primary or uncollected data. Most data can be categorized as either population data or sample data. Population data is an actual count of all occurrences, e.g., a listing of *all* fire calls for the year. Sample data is information gathered from only part of the population, e.g., a collection of evaluations on *some* of the public education presentations in the local schools. Population data are usually more accurate, but not always easy to get. Sample data can be very accurate as long as the selection process is followed correctly.

In order to use sample selection correctly, it is necessary to understand the concepts of **representation** and **randomization.** Representation refers to the fact that the members of the sample are "like" most of the members of the population. Their opinions or answers are typical of what one would get by asking other members the same questions. Since not everyone has the same opinions, however, it is necessary to ensure that all opinions have an equal chance of being expressed. This means that all of the different opinion groups in the population need to be represented. This is done through randomization, or **random sampling.**

Random sampling is the process of choosing members of the population to be part of the sample in such a way that every member has an equal chance of being selected. It is based on probability theory and aims to ensure that "each member of the population has equal probability of being selected." To sample a group because of size or time limitations, one must choose a random sample in order to avoid bias.

For example, if a survey is performed to determine what the "average" fire chief thinks about management training, but only graduates of the National fire Academy's Executive fire Officer program are surveyed, the sample is biased. These individuals share a common experience that will influence their opinions. Therefore, to achieve a random sample, which would truly represent fire chiefs across the board, chiefs from all parts of the country would have to be selected in a way that gave each one an equal chance of being selected.

A sample of members of the fire department would have to include fire fighters on each shift, of each rank, and in all staff assignments. In a combination department, both full-time and part-time personnel would have to be surveyed. Obviously, to find out how only the volunteers feel, they are the population of the study and only they need to be sampled. However, in this case one could not go down the list and select only those who are "active" regardless of how active is defined. One would have to decide how many to sample and then decide which random numbers to select, then take these numbers to the roster and select the sample without knowing who they might be. [Note: Random numbers can be generated by a computer or taken from the "Random Number Table" found in most statistics books.]

Using random numbers is a way of making sure the selection is not systematic in any way. Patterns in selecting samples have proven to result in patterns in results. Another method would be to use some means of *chance* to identify who is going to be chosen.

Activities of chance include things such as flipping a coin, rolling dice, or picking cards. Bingo and lotteries are considered games of chance. They are based on probability theory, which says each outcome has the same chance of occurring, all things being equal. (A one has an equal chance as a six of appearing when you roll a die.)

Examples of Random Selection

A volunteer department has forty fire fighters and five officers (fire chief + 44). The chief wants to interview ten of the members concerning their opinions of two types of turnout gear. He has a roster of the staff, numbered 1 through 44, in alphabetical order. The chief takes a deck of cards and decides that the Spades are lowest (Ace = 1, J = 11, Q = 12, K = 13), Diamonds are next lowest (Ace = 14, J = 24, Q = 25, K = 26), Hearts are next (Ace = 27, J = 37, Q = 38, K = 39), and Clubs are the highest (Ace = 40). He then sets aside all Clubs except Ace through 4, giving a total of 44 cards. He shuffles the deck and draws a total of ten cards, reshuffling between each draw. Each card has an equal chance of being drawn. The chief then matches the card number with the person on the roster with that number to select the interviewees. Thus, each member of the department has an equal chance of being selected. The chief has a random sample.

The assistant EMS coordinator in a small city (250,000) is ordered to evaluate the level of satisfaction of clients/patients over the past year. She looks at the following information:

12 months
12,624 runs
Average of 1,052 runs per month

By applying probability, the EMS coordinator determines that a completed sample of 400 responses is needed.

She decides to sample 800 clients who were served in the last 12 months because only 50 percent of surveys mailed are returned. In order to determine how many she needs to survey each month, she divides 800 by 12. [800/12 = 66 with a remainder of 8]. Therefore, she needs to select 66 from 4 months of the year and 67 from 8 months of the year.

In order to select the months for 66 and 67, she writes the numbers 1 through 12 on pieces of paper, folds them, puts them in a cup, and asks 4 people in the office to pick one. Those are the 4 months from which she will choose 66 people to survey. (Each month had an equal chance.) [Note: Initially she was going to roll a pair of dice four times to choose the 4 months, but she realized that it would be impossible to roll a 1. Therefore, January could not be selected.]

Dividing the average number of runs per month (1,052) by the number of clients to be surveyed that month (66 or 67) reveals that every fifteenth client needs to be surveyed. However, to simply start with the first client and then take the sixteenth, and then the thirty-first, creates the risk of having a pattern.

Therefore, the assistant coordinator borrows the Bingo calling machine from a neighboring volunteer fire department. She puts numbers B-1 through B-15 in the machine. She selects a number, records it, and replaces the ball. She does this twelve times, replacing the ball each time. She uses each number drawn as the starting number of the first case to select each month. After that, each fifteenth case is selected that month until 66 or 67 are chosen. This results in the following:

Month	*Starting Number*	*Number of Clients*
January	8	67
February	15	66
March	2	67
April	9	67
May	5	67
June	12	66
July	8	66
August	11	67
September	4	67
October	6	66
November	1	67
December	15	67

If the number of cases falls short for a particular month, the list for the month starts over. For example, February had only 970 cases. Case 966 was selected as sample number 63 for that month. Another 3 samples are needed. To maintain random sampling, the assistant coordinator counts the last four cases in February as the beginning of a new series of 15, selecting cases 11, 26, and 41 to provide the last three cases needed to fill the required number.

The assistant coordinator now has the sample list of clients who will be mailed the survey.

Both of these examples rely on chance to determine who will take part in the study. That is important in order to ensure that the sample is representative of the population.

Surveys

Surveys are often used to find out what people think. These are usually given to a random sample of the population that is the subject of the study. The survey questionnaire is the form with the questions on it that is mailed or handed out to the people in the survey. Constructing a survey is time-consuming and difficult without an appropriately trained staff; it is recommended that specialists be hired if this is the case. A few of the basic rules of survey development are presented here.

Some Basic Rules for Designing Questionnaires

1. Start with easy, nonthreatening, but necessary questions.
   Put difficult or threatening questions near the end.
   Never start with an open-ended question that requires a lot of writing.
   Demographic questions should usually go at the end of the questionnaire.
2. Start with general questions and move to more specific ones.
3. When gathering information on histories, such as jobs, shifts in the community, etc., collect it in chronological order, moving either backward or forward in time.

4. Place all questions on one topic in one part of the questionnaire. Do not skip from one subject to another. (For example, ask all questions about EMS before asking about public education activities.)
5. When switching topics, let the respondents know you are doing so.
6. Keep the questionnaire short by asking only the information that will be used. (See Chapter 8, Information Management, for a discussion of this.)
7. Remember that the respondent's interest in the topic influences the length of time he or she will be willing to spend on the project. The more interested respondents are the more time they are willing to spend. Do not try to get so much information that the respondent loses interest halfway through the questionnaire.
8. The appearance of survey forms has an impact on response. A questionnaire that looks easy to complete and is professionally designed and printed will get a higher response rate.
9. The name of the organization or individual conducting the study, a telephone and/or address at which to contact someone, and the name of the study should appear on the questionnaire.
10. Do not crowd questions; leave some white space on the paper. Make sure the survey is easy to read—use large type that is easy to read.
11. Number each question and letter subparts of a question to keep everything in order.
12. Do not split a question between pages.

Always make certain the questions are objective and unbiased. Ask: Which of the two nozzles currently in use do you prefer? Don't ask: Isn't the Eureka nozzle better than the FastBlade nozzle? A biased question steers the respondent toward a given answer, invalidating the entire survey.

Open-Ended versus Closed-Ended Questions

Closed-ended questions provide respondents with a fixed set of choices from which they select their answer. **Open-ended** questions allow respondents to write their own answers without prompting other than the question itself. Open-ended questions can provide much more information, but many people do not like to write and will not complete these questions very thoroughly.

Open-ended questions are also very difficult to code and interpret. It is hard to classify narrative statements into tables and other groupings. If everyone on the fire department were asked why they decided to become a fire fighter, the answers would range from one line such as "to help people" to long narratives about some inspiring incident. How could this information be classified and presented?

Another problem with open-ended questions is that they may yield more than one response. The researcher then must decide which is most important or how to organize all of the information. If the answers are unclear, the researcher might misinterpret the information.

For these reasons, it is generally better to use closed-ended questions in a survey. A place for comments at the end or a final question asking respondents to add anything they feel is important will allow them the opportunity to share additional information not requested in another part of the questionnaire.

A yes or no question is the simplest form of closed-ended question. The respondent must choose between two choices. This is appropriate for some pieces of information, but does not allow for variation in intensity.

It is a good idea to allow people to rate things by showing the degree to which they have an opinion. One way to do this is to use a Likert scale. Developed by Rensis Likert, this model consists of a scale of numbers, usually 1 through 5, 7, or 10[1]. There are at least two phrases serving as "anchors" at each end of the scale that tell the respondent what those numbers represent. The range enables respondents to identify the degree to which they agree with the descriptions on the anchors. Sometimes a middle number is also anchored, and sometimes all numbers are anchored. It is important that if all are anchored the distances are comparable.

Examples of Likert Scales

		Agree			Strongly Disagree	
Wrong:	The Eureka nozzle is easy to hold.	1	2	3	4	5

		Agree				Strongly Disagree
Right:	The Eureka nozzle is easy to hold.	1	2	3	4	5

		Excellent	Very Good	Good	Fair	Poor
Wrong:	The training session and material were:	1	2	3	4	5

		Excellent	Very Good	Good	Fair	Poor
Right:	The training session was:	1	2	3	4	5
	The material was:	1	2	3	4	5

		Inappropriate						Uninteresting
Wrong:	The training material was:	1	2	3	4	5	6	7

		Totally Inappropriate						Very Appropriate
Right:	The training material was:	1	2	3	4	5	6	7
		Interesting						Uninteresting
	The training material was:	1	2	3	4	5	6	7

Also, make sure each question is distinct. Don't mix two different ideas or components. In the second example of right and wrong Likert scales, the second wrong statement mixes the instructor's performance with the adequacy of the materials.

The third wrong example demonstrates what happens when the two anchors refer to different characteristics. It is important that each question ask only one thing. This is true for *all* qu[illegible]tions, not just those on the Likert scale. A yes or no question asking, W[illegible] the paramedic service prompt and courteous? for example, mixes up two performance criteria. If someone says no, was the service prompt but discourteous, slow but courteous, or both slow and discourteous?

Another popular form of questions is multiple choice. Multiple-choice questions must be **mutually exclusive** and **comprehensive.** Mutually exclusive means the different categories do not overlap. The following question is not mutually exclusive because someone could choose more than one category.

To how many calls did you respond last week?

_______ a. 1–10
_______ b. 10–20
_______ c. 20–30

If someone had responded to 10 calls, either (a) or (b) could be chosen.

Being comprehensive means that all of the possible answers are included in the choices. Because this is sometimes difficult to do, residents are often given "other" as the final choice. While it is good to catch the occasional unusual answer, if there are not enough categories provided the use of the other can result in the equivalent of an open-ended question. If that is the case, it is better to simply use an open-ended structure.

Interviewing

Personal interviews may be the best method of gathering detailed information that is not easy to classify and when the size of the sample is fairly small. Personal interviews involve the researcher developing an interview form to use in a one-to-one session with the subject or participant in the study.

These interview forms are generally one of two types: structured or guide. A **structured interview** form is one with very specific questions asked in a definite order. The interviewer never deviates from these questions. This is useful for obtaining specific information that is going to be compared in some way. For example, if there is a question about what happened at the scene of an incident, the investigating officer might construct the following structured interview:

1. What did you see when you arrived on the scene?
2. What did you do when you arrived? Go through everything step by step, identifying any orders you were given or that you gave.
3. Where was the third engine on the scene positioned?
4. What did the fire fighters on the third engine do once the apparatus was in place?

These questions do not ask for any impressions or perceptions, just for facts. The structured interview form will keep both the interviewer and the respondent on track and focused on the information at hand.

The **interview guide** is used in focused interviews when more perception and impression is needed. It is also useful when very little is known about the subject matter at hand. The interview guide uses some general questions and some statements or prompts to elicit responses. It encourages free discussion and allows the respondent to wander from the primary focus. The interviewer maintains the focus of the interview, using prompts or going to the next question if the topic switches or a major digression occurs. A sample interview guide that might be used to gather information on the perceived future of the fire department follows.

FOCUSED INTERVIEW GUIDE

What do you think the fire department will be like in the next 10 to 15 years?

Prompts: Personnel
Equipment
Responsibilities

How will changes in the community affect the operation of the department?

Prompts: Will the economic base change?
Is the area growing or declining?

[Note 1: Is the respondent aware of changes taking place in the community? Here the researcher would simply make a note on the basis of the responses.]

Interview Setting

Interviews should be conducted in private, comfortable areas where there will be no interruptions and where privacy can be assured. No telephones or station speakers should be located in the interview room, as they cause distractions. It is good to face the respondent in a cordial, nonthreatening manner so that facial expressions as well as changes in vocal tone can be observed. The interviewer must also be careful not to suggest answers by changes in his or her own expression or tone of voice.

Interviews should be scheduled in advance so that the respondents will know how much time will be demanded. Interviews should be practiced, then 10 to 15 minutes should be added to the practice time to schedule the interview's length. It is all right to take less time than planned, but taking more time can result in rushed answers or an incomplete interview. No interview should last for more than 1 hour.

Notes should be recorded on the interview form. These notes should be transferred or expanded as soon as possible so that the information is captured correctly. If possible, it may be advisable to tape record the interviews with the permission of the interviewee. This does mean, however, that tapes must be transcribed and then coded or otherwise organized. Because of this, it is often just as effective to take notes and write up each interview. In writing up an interview, however, the researcher must stay objective and record the comments made. Under no circumstances may the interviewer interpret or slant what the individual said. The purpose of an interview is to gather information, not to get people to say what the interviewers want to hear.

Remember that the rules and concerns for confidentiality hold for interviews as well as surveys. Don't say who said what unless they have given permission—in writing!

Existing Data Sources

Fire department records are one of the most readily accessible sources of data. Chapter 9 discusses the use of this information–how it is collected, processed, and made accessible. In addition, planning departments have a lot of data concerning population, land use, etc. Building departments have information concerning new and existing structures. Almost every department or agency in the fire department's service area has data relevant to its operation.

Good, professional relationships with the directors of these departments and agencies can facilitate access to data when it is needed by members of the fire service. (See Chapter 7, Interacting with Other Public Administrators.) When using data from other sources, realize that they also owe their clients confidentiality and may be forced to restrict access or to modify the form in which the data is made available. If one explains why the data are needed and how they will be used, the likelihood of getting useful information is far greater.

Needs assessments, evaluations, and other studies that these departments and agencies have conducted may also be of use to the fire department. For example, a study of social service needs of children that identifies the location of large concentrations of latchkey children indicates areas where unsupervised play takes place. In some larger cities this leads to arson and fires due to carelessness on the part of children. Studies by area agencies on aging identifying areas of the city with elderly and aging population provide indications of where there will be demands for EMS services. A local fire department can benefit from knowing what is happening in its community.

Libraries, especially research libraries associated with colleges and universities, usually have a number of data sources that may be of use. These sources tend to cover state and national areas as opposed to local information. Some of the information, such as the U.S. Census data, however, is broken down into regional comparisons and can provide relevant information. Some of the more well-known and frequently used sources are listed below.

Census of Governments, Washington, DC: U.S. Bureau of the Census

Census of the Population, Washington, DC: U.S. Bureau of the Census

County and City Data Book, Washington, DC: U.S. Bureau of the Census

The County Yearbook, Washington, DC: National Association of Counties and the International City Manager's Association

Municipal Yearbook, Washington, DC: International City Managers Association
The National Directory of State Agencies, Arlington, VA: Information Resources Press
Vital Statistics of the United States, Washington, DC: U.S. Government Printing Office
Statistical Abstracts of the United States, Washington, DC: U.S. Government Printing Office
U.S. Bureau of Labor Statistics, Washington, DC: U.S. Government Printing Office

ADDRESSING THE NEEDS IDENTIFIED

Once the needs assessment is complete, the next task begins. The information gathered now needs to be used appropriately.

Incorporating the findings of the needs assessment into the department's planning process is a primary use for this information. Goals and objectives should be set to meet the identified needs. In addition, the detailed information can be used to identify alternative approaches that may work best. This use of information and data is discussed in detail in Chapter 10, Planning.

It is important to share the results of a needs assessment. An assessment is a positive action taken in order to improve performance, not to identify targets for discipline.

Subordinate members of the department have a right to know the results of both community and department needs assessments. The fire fighters are going to be asked to focus their efforts at resolving the needs identified. Knowing exactly what the problem is, how severe it is, and how it was identified helps to focus attention and build commitment to solving the problem.

Information should be given openly and in a matter-of-fact manner. Holding back information breeds conjecture and rumor. When the information comes from the department administration, fire fighters develop a greater trust and feeling of respect and understanding for officers and administrators. This helps to dissipate adversarial relationships, improving department operations. It is best to provide the information in written form, supplemented if necessary by verbal explanation and detail. The written document will ensure that everyone gets the same basic information, again reducing discrepancies.

It is also important to inform superiors about the results of a study. This should be done in a prescheduled meeting where enough time is allowed to review the study and its findings carefully. It is also good to have a second meeting to answer any questions that arise after further study of the report. This will ensure that everyone has the same understanding of the information gathered.

Many of the studies conducted by outside consultants become public property as soon as the report is made to the governing entity or to the fire department. This is due to the requirements of the Freedom of Information Act and other federal regulations regarding public information. How the information is released to the public often affects how well it is received. This is where good relations with the media can make a difference. Sharing the results of a needs assessment with the media should be managed by a public information officer as discussed in Chapter 8, Information Management.

SUMMARY

Needs assessment and research techniques can improve the accuracy and objectivity of decisions made by fire departments. In addition, the appropriate analysis of data can help substantiate the decisions made. This is a technique that will be used increasingly in the fire service. Familiarity comes best with practice. As a result, fire officers are encouraged to follow the steps provided and practice the techniques demonstrated in order to make better use of information and become more familiar and comfortable with the techniques.

ENDNOTE

1. For a discussion of the development, construction, and use of a Likert scale as a measurement method, see Earl R. Babbie, *The Practice of Social Research,* 2nd ed. (Belmont, CA: Wadsworth Publishing Co., 1979), pp. 409–412.

REFERENCES

Monette, Duane R., T. J. Sullivan, and C. R. DeJong. 1986. *Applied Social Research.* New York, NY: Holt, Rinehart and Winston.

Patton, Michael Q. 1990. *Qualitative Evaluation and Research Methods.* 2nd ed. Newbury Park, London: Sage Publications.

Rossi, P. H., and H. E. Freeman. 1982. 2nd ed. *Evaluation: A Systematic Approach.* Beverly Hills, CA: Sage Publications.

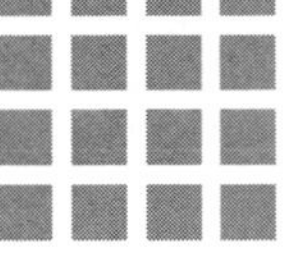

CHAPTER 10

Planning

Planning is a management tool. When used by fire department administrators, planning delineates a variety of line and staff activities to lead the fire department to operate at a higher level. Depending on the size or administrative structure of the department, the planning function will address basic pre-incident evaluation, resource utilization, equipment selection, and future service demands of the community. Once the department's emergency response plan is developed and approved, it will often be incorporated into the community's emergency response master plan. Master plans are generated by local public administrators and elected officials who actively pursue the development of a comprehensive emergency response plan that is geared toward coordinating the efforts of all local agencies. The hazardous materials response plan mandated by SARA Title III is an example of such a plan.

Although planning is recognized as an effective management tool, some fire department administrators limit its use to operational emergency response issues and policies only. However, excluding budget preparation, resource allocation, and organizational development from departmental planning reduces the scope and effectiveness of the agency. Nonemergency functions become secondary and are often administrated by other local government representatives. The most effective planning is comprehensive, incorporating emergency response planning into the overall fire department operations plan. It is generally acknowledged that the desired result of any emergency response agency is the successful mitigation of an emergency incident; however, planning should transcend emergency response operations and include all functions and divisions within the organization.

The fire administrator must strive to ensure that the planning process produces a comprehensive document that provides operational and administrative direction for current and future department activities. The planning document answers the **who, what, when, where,** and **how** questions that can be used to implement, manage, and direct the organization's resources and activities.

WHAT IS PLANNING?

Planning is a managerial function that establishes specific goals and objectives aimed at achieving a predetermined outcome.

Planning will (1) assist in accomplishing the mission of the organization, (2) identify the organization's direction, (3) provide a written guideline for management, and (4) define the type and level of evaluation needed to determine whether the goal has been achieved.

Planning will also affect budget preparation and resource allocation. Without a clear plan of action, resource allocations may be inadequate, causing programs to be underfunded, and prevent the organization from operating at optimal levels. In addition, the plan can be used to justify budget requests, as it identifies exactly what resources are needed, how they will be used, and what will be accomplished. The plan should become a supporting document for the budget process. (Conversely, plans that require excessive resources are not beneficial, as they cannot be carried out.)

Once the fire department plan is finalized, it becomes part of the comprehensive community plan. This emergency response plan integrates the fire department plan with other emergency response agencies such as police, emergency management, transportation, utility, transit authority, and emergency medical services. The end result is a cohesive municipal emergency response plan that will provide direction in time of emergency.

The community plan, equally important yet often overlooked by fire departments, is the organizational/development plan for the community. It considers planned growth, annexation, infrastructure changes, upgrades, expansion, and road network improvement and development. Fire service concerns and needs should be considered when elected officials and public administrators evaluate changes in

FIGURE 10-1 Planning is a management tool that utilizes input from a variety of sources.

the infrastructure or makeup of the community. While it is clear that fire protection is a consideration when discussing annexation, it is not as easily or quickly recognized that the fire department should also be consulted when the road department is prioritizing its projects. The fire department needs to be aware of planned road closings to develop alternative routes in advance of an emergency situation. The fire department also needs to have the opportunity to provide input concerning use of access corridors. For example, if the volume of calls is increasing in one neighborhood and there is only one access road leading in and out of the area, the fire department must document service delivery difficulties and work with the road department to try to develop a secondary corridor. Thus, when approached correctly, the planning process not only maps out the future of the department, but it also integrates the fire department with the rest of the community.

Mission Statement

An organizational plan is formulated by creating and adopting a good mission statement. It is difficult to plan activities when it is not clear what the goal is or why it is being pursued. Although many fire departments have adopted mission statements, planning can be frustrating since some of these statements are generic and do not address local problems or accurately identify the purpose and responsibilities of the fire department. While statements such as "to save lives and property" are certainly the basis of American fire service tradition, they tend to be too vague and difficult to measure. In other instances, the level of service delivery addressed in the mission statement is unrealistic and beyond the capabilities of the fire department and even the local community.

The National Fire Academy offers the following definition: "A mission statement is a brief but meaningful statement of the reasons for the department's existence, the ways in which it accomplishes its purposes, and the values that drive the organization."[1] In other words, the mission statement should focus the organization's activities and clearly identify the goal it strives to achieve.

The National Fire Academy has also identified seven criteria for evaluating a mission statement. It must:

1. Clearly identify what the purpose of the department is, in terms of what needs the department aspires to fulfill and how this will be accomplished
2. Reflect the values, beliefs, philosophy, and organizational culture of the department
3. Be broad enough to permit flexibility in implementation but not so broad as to distort the focus of the mission
4. Reflect the aspirations of the department and its members, but be reasonably achievable
5. Be clear and understandable to all personnel
6. Be brief enough for most department personnel to remember
7. Serve as the basis for daily decisions and actions by all members

The mission statement should be reviewed every 2 to 3 years, using the above criteria as a base for evaluation. Even if no changes are made to the mission statement, a thorough review can revive the statement and the department's dedication to it. The order in which the various activities are listed in a mission statement can indicate their relative importance to the department. A sample mission statement follows.

> *To prevent fires through fire safety education, code enforcement, preplanning, and fire investigation. To provide emergency medical service, fire suppression, and hazardous materials spill mitigation to the community. To provide the highest level of service feasible for the level of resources allocated to the organization.*

Any organizational plan must begin with an understanding of the organization's mission. How can it be determined whether the mission statement is best suited for the type of organization and service it provides? One sure method is to critique it. Here is a sampling of the questions to ask when trying to assess a mission statement's full impact:

1. Does the mission statement fit the organizational structure?
   - *If not, what organizational structure best fits the mission statement?*
2. What service(s) does the organization provide?
   - *Examples: fire safety education, EMS, suppression, hazardous materials response, blood pressure screening, fire safety inspections, hydrant servicing, other public services*
3. Is the identification "fire department" most appropriate?
   - *If not, is a name change feasible?*
4. Is the statement too generic in that it provides for all contingencies regardless of capabilities and resources?
5. Should the mission statement be completely revised to better reflect the current status and structure of the organization?

Evaluating the mission statement offers an excellent opportunity to involve other officers and supervisors in the administrative process. Their input may help to identify other issues and concerns that should be addressed in the mission statement. In order to maintain some level of uniformity throughout this evaluation process, the use of a Likert scale rating system may prove valuable. This evaluative tool, as discussed in Chapter 9, Needs Assessment, provides each participant with a standard form of measure, making the rating procedure much easier.

FIGURE 10-2 Mission Statement Evaluation Form

Question	*Degree of Agreement* *Weak*				*Strong*
1. The mission statement fits the organizational structure.	1	2	3	4	5
2. The services of the organization have changed since adoption of the current mission statement.	1	2	3	4	5
3. The name of the department ("fire department") most appropriately describes what we do.	1	2	3	4	5
4. The statement is too generic in that it provides for all contingencies regardless of capabilities and resources.	1	2	3	4	5
5. The mission statement should be completely revised to better reflect the current status and structure of the organization.	1	2	3	4	5

If it is determined that the statement is not appropriate for the organization, a new and more relevant statement should be drafted as soon as possible. The best approach is to form a task force made up of personnel from all ranks and divisions. The assignment should be explained in detail and a specific time line for completion should be established. Checkpoints for administrative review must be built into the project. This ongoing communication will prevent the task force from pursuing an element or issue that is inconsistent with the policies or needs of the local community or beyond the resources of the fire department. Depending on the local government structure, the final mission statement may have to receive formal approval from the governing entity outside the fire department. Upon receiving such approval, the adoption of the mission statement deserves a formal presentation to all personnel. In independently organized departments the mission statement should be adopted by all personnel, and a formal presentation should still be made.

While some criticize this process as being too involved, experience has shown that if the mission statement is not clear, appropriate, and accepted, the planning process is severely hampered because it does not have a clear direction. Just as it is important to identify the questions in a needs assessment, it is critical that the mission be identified for a fire department before beginning the planning process. Time spent here will be saved later.

Involving officers and fire fighters is important because participating in writing "their" department mission statement builds true ownership of the department. When fire department members have the opportunity to determine the future of their organization, they develop a sense of pride and ownership. This creates high morale. A more dedicated and motivated workforce then performs higher-quality customer service.

Issues that Influence Planning

It is important for fire service officers to become aware of local issues that will influence the planning process. A plan that is written in isolation, without regard to real-life restrictions or problems, is unusable. The following discussion first identifies issues and then discusses three techniques to assist in identifying goals and objectives: gap analysis, force field technique, and needs assessment. These techniques clarify the goals and objectives so that the plan can be more effective.

The environment in which fire service administrators must function and manage often plays a significant role in the development of the department plan. According to S. R. Michael, "The decision environment refers to conditions that have a bearing on what a manager is actually able to implement."[2] This environment includes both internal and external conditions. **External** conditions are those conditions or circumstances that exist outside the organization. Henry Mintzberg[3] classified external influences as major disasters (such as floods, tornadoes, and war), economic conditions, cultural trends, political considerations, technological developments, and community socioeconomic status, all of which may affect the way a manager approaches the planning process. Even though little can be done by the department to control these events, the prudent administrator must provide a contingency for such occurrences.

The **internal** environment, according to Herbert Simon,[4] includes personnel issues, budget preparation or allocations, labor-management concerns, and federal, state, or municipal laws. These are ongoing issues that involve department personnel and management on a daily basis. Certain internal aspects of the organization are fixed and must be considered when developing a feasible plan.

Often external and internal environmental influences are beyond the control of the administrator; however, both must be dealt with in organizational planning. These environmental factors are often termed **boundaries,** as they identify the limits placed on the plan.

Short-Term, Long-Term, and Strategic Planning

A fire officer can plan activities for a day, a month, or a year. A department can plan for a day, a month, a year, a decade, or a century. Each of these plans will have a different focus and different details. One can expect a different amount of adaptation or change during a plan's implementation based on the complexity of the plan. One of the first decisions to be made when constructing a fire department plan is setting the time frame for achieving goals and measuring progress. This is accomplished by designating specific periods of time (usually in months or years) in which the project is to be initiated or completed.

It is important to understand that any time-line system is not cast in concrete and can be modified as necessary. The plan's time line should contain checkpoints to review progress to date. If changes need to be incorporated, the time line should be adjusted accordingly. Internal and external factors may change and influence the department's ability to implement the plan as scheduled. Therefore, the plan must be flexible enough to respond to these influences. Since the plan will contain a number of actions that should take place at various times throughout execution of the plan, it will be necessary to determine: (1) the required use of specific equipment or services, (2) the administrative attention required to initiate such action, (3) the approximate cost involved, and (4) the financial resources available currently and projected in the future.

Available resources, whether they are financial, administrative, equipment, or some other types, must be allocated realistically. For example, a project that requires 20 hours a week of officer time during its first 3 weeks might not be able to begin if the officers' time is currently assigned elsewhere. Meeting the project's supervisory condition would require a time-line extension or reassigning officer time. Thus, project plans and time lines must be coordinated with ongoing operations. This results in comprehensive department operations plans for a specific time period.

In order to determine specific time lines, three time intervals—short, long, and strategic—are established. **Short-term** planning refers to a period of 1 year or less and focuses on activities that will take place during the current calendar or fiscal year. Quite often, short-term planning is tied to the operating budget adopted to accommodate the current year's expenditures. Items such as small equipment replacement, fire safety education programs, station and equipment maintenance, and training programs are generally placed in the short-term planning period. The annual operating plan of the department is considered in the short-range time period.

When preparing a short-term plan, the administrator must exercise a degree of caution in terms of what can actually be accomplished as compared to what is the desired level of accomplishment. Reasonable expectations must drive the plan. A realistic approach to short-term planning includes only those issues that will be addressed during the period of time with the resources available. Don't include projects or personnel that require unbudgeted funding or significant re-allocation of resources unless it is reasonable to consider that resources can be made available.

Writing unattainable plans and goals demonstrates a lack of understanding of the organization's capabilities. When done repeatedly, department morale suffers as fire fighters and officers continuously see promises made and never kept. In addition, elected officials and senior public administrators begin to doubt the management ability of a fire chief who never plans accurately. Although planning requires some guesswork, short-term plans are the most accurate.

Long-term planning covers a time period from 2 to 5 years. It is usually grounded on predictions or goals that provide the future direction of the department. Examples of long-term planning include: (1) station location or re-location, (2) increased staffing, (3) apparatus replacement, (4) new programs or divisions, (5) major shifts in service provisions, and even (6) decreased staffing. The long-term plan incorporates the needs identified in a department or community needs assessment. The time line should reflect the priorities assigned. An example of an appropriate component of a long-range plan would be the adoption and implementation of NFPA 1500, *Standard on Fire Department Occupational Safety and Health Program.* Implementing the requirements of this standard will take time because of the wide range of safety aspects it contains.

The administrator must not consider the long-term plan as a rigid course of action that cannot be modified. Once the plan has been accepted, the manager should evaluate the progress of each element and be prepared to adjust any portion of the plan to compensate for external or internal changes. For example, the mayor of a southwestern community announces that the latest census report indicates a 19 percent decline in population during the past 10 years. This was due to significant business reversals and plant closings that occurred during the decade, causing citizens to move to other areas of the country for employment. The fire department's long-range plan calls for a 6 percent increase in station personnel each year. Therefore, the fire administrator must re-assess the long-range plan to determine if that level of manning is still required in all stations or only certain ones. Also, consideration must be made as to whether the action is fiscally prudent since a decrease in tax revenue is anticipated.

In this case, being prepared for all emergency contingencies may no longer be feasible in this community. A *modified* long-term plan would indicate some downsizing of the department to reflect the impact of decreased revenues.

The overall direction of the department is identified in the **strategic plan.** This plan should reflect the goals and administrative aspirations of the department for the next 5 to 15 years. Thus, any aspects of NFPA 1500 requiring more than 5 years should be incorporated here. In addition, because the strategic plan reflects the long-term philosophy of the department, the commitment to safety exemplified by the standard should be contained in the plan as well. It should also reflect the anticipated growth or decline of the service area, changes in the environment, and anticipated new services for the future.

Comprehensive community emergency response plans are considered strategic because they identify weaknesses in response capability and steps to take to mitigate them. In addition, these plans are timeless because they remain in effect until modified and updated. Comprehensive community emergency response plans are a combination of short-range and strategic planning because they detail specific actions to take in an emergency, designate strategic activities for the department and community, and validate commitments to improving overall safety and response preparedness.

Good strategic planning relies on accurate information and identifies an acceptable level of risk for the fire department and the community it serves. It should not be restricted by present department policies or fire service tradition. Creative thinking must be encouraged of all participants developing the plan. It is often said that the strategic plan is often not achievable in terms of specific elements. However, it should be considered a guide to the evolution of the fire service as the service prepares for the challenges of the future.

PLANNING TECHNIQUES

In order to produce a planning document that will serve as an effective management tool, fire administrators must first evaluate the current status of department activities. Any organization, regardless of size, response capability, or resources, will dominate in certain areas of service and do poorly in others. No fire department can excel at everything it attempts. The key to increasing the level of success is to identify the activities or service delivery systems judged to be superior and those that are considered good, moderately successful, or even poor. For the final analysis to be as objective as possible, the task of gathering the information necessary to assess these activities and services should be coordinated with all department administrative and line officers.

The first step to improving services is open and accurate identification of problems. Failing to identify or admit to real problems ensures the defeat of plans to rectify the cause of poor service delivery. Problem identification is a critical process and requires careful attention and effort on the part of department administrators.

Root versus Symptomatic Problems

As the administrator begins to plan, the first task is identifying root and **symptomatic** problems. In some instances, situations or conditions exist that mask the true root of the problem. Often, a symptomatic problem is the first to be recognized and identified as a problem. However, what is recognized may actually be the symptom or result of another underlying problem. These circumstances may go unrecognized and be misunderstood for long periods of time. Some problems are symptomatic of much more serious root problems.

If the administration attempts to address the visible symptomatic problem without first identifying the true or root problem, the adverse condition will not go away. At best, the root problem manifests itself in other symptoms. Too often, failure to address the root problem causes it to grow larger and adversely affect the entire department. Therefore, it is critical that the problem be carefully examined. Plans to remedy dysfunctional activities in the department must be aimed at the root.

For example, a department sustained a significant increase in the number of workers' compensation claims filed. A review of these claims showed that the majority were back injuries. The chief decided to institute a physical training program to improve back strength having identified poor physical condition as the reason for injury. A year later, the frequency of injuries had not decreased. Reexamining the situation, the chief discovered that the root of the problem was that the training module on proper lifting techniques had been eliminated from the annual training schedule. Thus, the true root problem was lack of training.

Other examples of root versus symptomatic problems:

1. Excessive tardiness—may be symptomatic of a root problem of lack of motivation
2. Smoke inhalation injuries—may be symptomatic of such root

problems as inadequate training, unclear standard operating procedures, or outdated equipment

3. Low water pressure to fight structure fires—may be symptomatic of such root problems as pumper malfunctions, inadequate municipal water grid system, unqualified pump operators, or insufficient water supply hose line

Attempts to solve a symptomatic problem without understanding the root problem will be futile. Planning begins with problem solving, and problem solving means recognizing and correcting the root problem first.

Methods of assessing a program, unit, or department are called planning techniques. The goal of planning is to accomplish the mission, and planning techniques link the function with the mission. Needs assessment is one of the most comprehensive methods of identifying problems in ways that facilitate planned attempts to solve them. There are other useful planning techniques, although they are not as comprehensive. When root problems are known or specific goals adopted, however, these techniques are useful for planning solutions to problems. The same techniques can be used by officers to create a career development plan.

Gap Analysis

The purpose of the gap analysis system is to develop a reasonable understanding of where the department is now compared to its objectives. In order to determine the status of the organization, a performance assessment should be conducted and the findings compared to present expectations as well as the physical and fiscal capabilities of the unit.

Once the current status of the program or organization has been determined, and the desired level of accomplishment has been clearly identified, the gap between the two is obvious. The next step is to ask two questions:

1. What does the organization need to do?
2. How can the task be accomplished?

In other words: How can the gap be closed? The answers to these questions become the foundation for the planning process. This information can also be used as a means of monitoring the progress of the planning activity.

For example, a fire department currently inspects 90 percent of public and private schools and 80 percent of the places of public assembly annually and wants to conduct annual inspections of 100 percent of both types of structures. These inspections are utilized for *both* preplanning activities and code enforcement. A gap analysis reveals the following:

Current Status Annual Inspection	*Gap Analysis Results*	*Goal Annual Inspection*
90% schools	I. Certify another fire inspector A. Select candidate B. Secure training C. Reallocate time [D. For part-paid departments: secure funding for extra hours] [E. Fund certification training]	100% schools
80% places of public assembly	II. Set priority for these structures and assign inspections accordingly A. Create a new assignment system	100% places of public assembly

Now it is clear what needs to be done to close the gap and meet the goal. This foundation may be used to close the gap immediately if resources are available, or to develop a plan to meet the goals gradually.

Force Field Analysis

Another administrative evaluation tool is the force field analysis system. This method is based on identifying the driving and restraining forces that might impact the success or failure of the organization in meeting the goal, or closing the gap. For example, a department may face external environmental issues, such as population erosion or economic downturn, that have a negative effect on the organization. On the other hand, passage of new local fire safety legislation will drive the need for more inspection and enforcement personnel.

The force field analysis method will assist the administrator in identifying the current status of the organization and those forces that may drive or restrain it both currently and in the future. The force field analysis is a refinement of the gap analysis, as it begins to identify and define resources and opposition to preferred goals or objectives. Each of these techniques can be used to prepare personal development plans as well as organizational plans.

FIGURE 10-3 Personal and Organizational Use of Gap and Force Field Analyses to Achieve Goals

Personal: An engine company captain desires to become a fire investigator.

Gap Analysis: Current position: Engine company officer
Desired position: Fire investigator

Analysis: For this fire department, the requirements are 100 hours of fire investigation training and 200 hours of law enforcement training. The captain has completed the 100-hour fire investigation training. Thus, the gap consists of 200 hours of law enforcement training.

Force Field Analysis:

Driving forces:

- Long-range plan calls for increasing the number of fire investigators
- Officer has a health problem that may begin to interfere with performance of engine company duties

Restraining forces:

- Department has no training funds left to pay for course
- Course is held at the state law enforcement academy, located 150 miles from the department

At this point, the captain can begin to explore alternatives and plan how to achieve the desired goal.

FIGURE 10-3 (continued)

Organizational: The fire department has a goal to create a fire investigation unit.

Gap Analysis: Current status: No fire investigation unit; two certified fire investigators on staff
Desired status: Operational, fully staffed fire investigation unit

Analysis: Need three more certified fire investigators as well as organizational structure change to include the new unit.

Force Field Analysis:

Driving forces:
- Goal approved in long-range plan
- Department presently contracts for fire investigation services
- State fire academy offers the 100-hour fire investigation course on a quarterly basis

Restraining forces:
- Department budget has been cut 4.5 percent
- Costs of the fire investigation and law enforcement academy courses have just been increased 10 percent
- Law enforcement course is only offered semiannually

NEEDS ASSESSMENT

The most comprehensive method of systematic organizational analysis is the use of a needs assessment, which is discussed in detail in Chapter 9. To review briefly, a needs assessment is employed when both the problems and needs of the organization require clear definition and attention. This technique identifies the relative severity of needs of the organization so that priorities can be set. A needs assessment can be performed on a community, an organization, management, equipment, or an individual. The needs assessment further refines the gap analysis and force field analysis techniques.

The specific steps are repeated here to illustrate how needs assessment is a refinement of the force field technique. The examples discussed for each step relate the use of needs assessments in planning.

Procedures for Conducting a Needs Assessment

1. *Define the focus of the assessment.* Set the parameters for the project. For example, is it to cover the entire department or only a specific bureau or division?
2. *Collect data.* Assemble all existing information and data on the subject of the study. Be sure to include biased along with objective information from as many sources as possible. Quite often, personal opinions are disregarded in favor of documented information. This is not the correct approach. All available data should be gathered to ensure that the study is as comprehensive as possible. (When biased information is used, identify the type and degree of bias, then use the information constructively.)
3. *Generate data.* In some instances, the information necessary may not be available. Therefore, it will be necessary to collect the primary data through research surveys and focused interviews. An example of this technique is to use a focused interview with all of the line officers to determine whether the department's emergency response standard operating procedures are being incorporated into engine company training programs. Other data collection activities may include the charting of department responses by geographic area by time of day and day of week.
4. *Organize data.* All the data collected must be organized into a usable format. If the administrator is attempting to identify stations with the greatest number of fire fighter injuries, simply listing the total number of department personnel injuries in random sequence will not provide the information necessary to answer the question. The data should therefore be organized by shift within station to provide an accurate status of injuries. This reorganization of data may lead to the discovery that certain injuries tend to occur more frequently on certain shifts or station apparatus. If this is the case, it then is recognized as a symptomatic problem that the administrator can use to identify and correct the root problem.

5. *Analyze the information.* The data collected must be analyzed appropriately. One method is the **compare and contrast technique.** This is a system of identifying the similarities, differences, trends, and patterns in information.

   For example, in a management needs assessment of a combination fire department, the issue of insufficient opportunities for off-site fire fighter training is addressed. The career fire fighters identify the problem as favoritism toward the paid on-call fire fighters, whereas the paid on-call fire fighters identify the problem as a lack of formal communication from the training officer and the fire chief.

   Additional data may be needed in order to identify the root problem. In this case, an investigation revealed that the paid on-call fire fighters were informed of training opportunities; however, the deadlines for application had often passed. The career fire fighters, in most instances, were aware of the opportunities, but very few had applied to attend.
6. *Interpret data.* Once the data have been analyzed, the results must be translated into a form useful to the organization. It must relate to the results of the study.
7. *Determine needs.* After the data have been interpreted and the problem(s) identified, it is necessary to plan how the problems will be rectified.
8. *Set priorities.* The final step in the needs assessment is to arrange the priority of needs, from highest to lowest, based on which are most critical to provide quality service. The activities necessary to address the needs of highest priority are then scheduled.

A needs assessment is a systematic approach to identifying problems and preparing solutions. To be successful, a plan based on the needs assessment must be systematically employed, and the information must be assessed critically and objectively examined. Wishes must be transformed into goals and objectives.

Writing Effective Organizational Goals and Objectives

Once the organizational problems or needs and priorities have been identified, the task of writing specific goals and corresponding objectives can begin. A review of the mission statement coupled with a clear understanding of problems provides the necessary foundation for developing the goals and objectives.

Goals are clear, broad statements about the direction of the organization and what is to be accomplished. Goal statements generally describe anticipated outcomes that are tied to the organization's mission statement. In most situations, goals will emanate from the top down in the organization.

Caution should be exercised to avoid ambiguity when writing goal statements. Stated goals should not be too broad for the resources available or written in such a way that it is not feasible to achieve the goal. There should always be a method for measuring the success or failure of the goal. If there is no fiscal or bureaucratic accountability built into the goal statement, it should be reworked to include a realistic method of evaluation. Did the department accomplish what it set out to do? How was the success or failure determined? The fire administrator must be prepared to answer these questions for the elected officials or the administrative manager of the municipality. In many instances the goal will be tied to revenue allocations and will require a certain level of accountability.

Examples of Goal Statements

Goal A in each set is an example of an inappropriate goal statement. Goal B in each set presents the same goal in a more usable format.

1A. Save lives and property.
1B. Respond to all fires or other emergencies occurring in the community and protect lives and property within the limits of the fire department resources.
2A. Handle all hazardous materials incidents.
2B. Respond to all hazardous materials incidents occurring in the community and contain the hazardous substance, protect the citizens of the community, and protect the environment within the resources and capabilities of the fire department.
3A. Provide parking for all fire fighters and visitors.
3B. Provide parking at each station for on-duty fire fighters and other citizens who may visit the station.

Objectives are statements identifying the methods or means chosen to achieve the goal. More than one objective may be applied to each goal to provide all of the necessary elements. Objectives must be clear, time-specific, and measurable. The process of setting objectives should involve those personnel responsible for meeting them so that the overall goal and objectives are understood by all.

Good objectives are: (1) written, (2) challenging but realistic, (3) regularly updated, and (4) assigned priorities.

Example of Goal and Corresponding Objectives

Goal: Respond to all fires and protect lives and property within the limits of the fire department's resources.

1. Prepare a list of all department apparatus and equipment within 30 days.
2. Prepare a list of all personnel and current levels of training for each within 15 days.
3. Develop standard operating guidelines or procedures (SOGs or SOPs) and appropriate training materials within 4 months for all department emergency response operations based on current apparatus, equipment, and personnel training levels.
4. Schedule department training programs to present the new standard operating guidelines and conduct simulated response training. To be completed within 6 months.
5. Inspect all commercial, industrial, educational, and multifamily occupancies within 6 months and prepare hazard analysis reports for each.
6. Develop specific emergency contingency plans for all target occupancies identified during the inspection phase. This action will be finalized within 4 months following the completion of the inspection phase.
7. Conduct officer training program focused on the new SOG (SOP) and emergency response plans within 1 month following completion of both.

Action Plan

Once the goals and objectives have been written and agreed on by all those responsible for their implementation, the **action plan** should be prepared. This portion of the planning process describes how the method to achieve the objectives is selected and initiated. In narrative form, it details the steps to be taken, the personnel involved, and the deadlines for each objective within the time frame specified.

Sample Action Plan

The fire department will institute a new emergency response policy within the next 12 months. Initial data collected will include information on the current apparatus and equipment, personnel training, and occupancy hazards in the community. The personnel data will be supplied by the senior captain, and the apparatus and equipment data will be gathered by the first assistant chief. Occupancy inspections are the responsibility of the lieutenant in charge of fire prevention, and the target hazard analysis will be performed jointly by the lieutenant and the fire chief. Training for officer and other emergency response personnel will be conducted by the assistant chief in charge of training. All training materials will be developed by the fire chief, the assistant chief in charge of training, and the lieutenant in charge of fire prevention. Regular meetings are to be scheduled the first Wednesday of each month in order to update all concerned on the progress to date. Implementation of the new policy will take place 12 months from the date of adoption of this plan.

The action plan is often prepared using forms that identify the goal, objectives, specific tasks, and resources required. These forms correlate with the narrative description of the plan presented above. Either is appropriate for fire department use in plan implementation. The forms list the individual actions required and the resources allocated. The narrative is a more integrated overview that relates the various activities to one another. For this reason, both may be beneficial.

A separate page is usually used for each objective. In addition, the completion of each objective and each goal is assigned to an appropriate individual in the organization.

Sample Action Plan Form

Goal: Initiate a new emergency response policy within 12 months.

Objective: Prepare a list of all department apparatus and equipment within 30 days.

Tasks	*Time*	*Resources* *Personnel*	*Equipment*
Evaluate equipment condition	14 days	Senior station officer	Standard forms
Document station by station equipment availability	7 days	Senior station officer	Standard forms
Consolidate equipment list and evaluations	7 days	1st Assistant Chief	Dept. computer system

Implementation of the Plan of Action

Several steps must taken to implement the plan of action. Specific tasks will be assigned to the responsible individuals to ensure that each objective is met. The administrative and capital costs of implementing the program must be monitored. Once the program is in place, monitoring the program will require minimal resources.

SUMMARY

Planning is a critical management tool that will help fire administrators prepare for the future and remain aware of difficulties providing services. Planning can provide the foundation for other administrative duties and identify what needs to be done and how. Planning also supports accountability as a mechanism to determine when objectives are not met and why. When planning techniques are used, the root and system problems that contribute to unmet objectives can be identified and amended, reducing frustration. In short, planning is the most integrative management function, tying the various aspects of managerial duties together. It is a tool that, properly used, can enhance the future and daily operations of the department. Failure to plan can catch a department unprepared to meet the needs of today and the challenges of tomorrow.

SAMPLE PROBLEMS

Problem 1[5]

Fire Department Mission Statement: The mission of the Municipal Fire Department is to continually evaluate the quality of the service delivery system and ensure that public safety and emergency services are provided in a caring and effective manner.

In the past 6 weeks the fire department has experienced a serious EMS response problem. On several occasions, EMS units responding to medical emergencies were forced to terminate the run and request assistance from another station in the community. This situation has resulted in significant delays in emergency transport. According to the fire chief, vehicle maintenance records show that these emergency runs were terminated due to an excessive number of flat tires. In order to alleviate the problem on a temporary basis, the mayor has directed the police chief to have on-duty police officers respond to medical emergencies along with fire department EMS units. The problem is that not all police department personnel have been trained to provide basic life support assistance.

1. Symptomatic problems:
   (a) Unwanted termination of EMS responses
   (b) delay in transport of patient
   (c) all police officers are not trained to BLS level
2. Root problem: Excessive number of flat tires on EMS response vehicles
3. Goal statement: The fire department will institute a new policy concerning tire maintenance and purchasing aimed at eliminating the unwanted termination of EMS calls and delays in patient transport.
4. Objectives:
   (a) Within the next 10 days, analyze the frequency of problems reported with the tires currently in use
   (b) Each EMS response unit driver to make a visual inspection of the tires on that vehicle at the beginning of each shift

(c) Institute a program immediately to monitor the causes of flat tires and classify each incident by type of damage
(d) Contact other regional fire departments in the next 2 weeks to identify which brand of tire they are using and the rate of failure
(e) Select the most dependable tire and purchase new replacement tires for all EMS vehicles by the end of the month

Problem 2

The following problem demonstrates how the techniques presented need to build on one another. They are not to be used in isolation.

This problem demonstrates how the planning process is applied to Needs Assessment 2 in Chapter 9. The mission of the fire department used in this case study is presented below. Note how the goals, which incorporate the prioritized needs, are consistent with the mission.

Fire Department Mission Statement: To prevent fires through fire safety education, code enforcement, preplanning, and fire investigation. To provide emergency medical service, fire suppression, and hazardous materials spill mitigation to the community. To provide the highest level of service feasible for the level of resources allocated to the organization.

Restatement of the Prioritized Needs
Identified and Prioritized from Chapter 9

1. There is a need to formally notify fire fighter/paramedics of the probability of requiring ALS when responding to patients in Districts 5, 8, and 13.
2. There is a need to inform the leaders of the city and of the immigrant communities about the high proportion of severe accident and illness calls.
3. There is a need for education about the proper use of the EMS system.
4. There is a need to introduce more preventive care practices.

In response to the needs assessment, the fire department sets three goals to be met in either its short-range or long-range plans:

Goal I: To present a culturally appropriate fire safety program to immigrant communities in Districts 5, 8, and 13 in order to decrease the higher-than-average number of fires in these neighborhoods.

Goal II: To provide high-quality Advanced Life Support Emergency Medical Service care whenever called for by residents of the communities.

Goal III: To reduce the abnormally high number of EMS calls from residents who require ALS care.

Once these goals are adopted, an action plan is developed in order to define objectives and identify specific tasks for accomplishing each. The following action plan forms present the chart format for these plans. As you examine them, ask yourself whether they are short- or long-range goals and whether they are viable in terms of the probable level of commitment required.

Goal I: To present a culturally appropriate fire safety program to immigrant communities in Districts 5, 8, and 13 in order to decrease the higher-than-average number of fires in these neighborhoods.

Objective I-1: Contact community service agencies in the city that work with the identified immigrant communities and secure their cooperation in this project within the next 3 months.

Tasks	*Time**	*Resources*	
		Personnel	*Equipment*
Identify relevant agencies and contacts	14 days	Capt. Herzog	New forms
Host an information meeting for agency representatives findings	45 days	Pub. ed. officer, Capt. Herzog, fire chief	Training room, copies of needs assessment
Identify liaisons and task force members to pursue the project	60 days	Pub. ed. officer, agency contacts	
Make the task force official	90 days	Agency directors, Mayor	

*Time is measured from the inception of the project. Thus, the second task is completed 30 days after the completion of the first task, which was completed 14 days after the beginning of the project.

Goal I: To present a culturally appropriate fire safety program to immigrant communities in District 5, 8, and 13 in order to decrease the higher-than-average number of fires in these neighborhoods.

Objective I-2: Develop a culturally appropriate fire prevention and safety program to be delivered in the identified immigrant communities.

Tasks	*Time**	*Resources*	
		Personnel	*Equipment*
Explain fire prevention programs to members of the task force	14 days	Pub. ed. officer	Copies of materials
Adapt the programs to cultural differences	175 days	Agency reps., pub. ed. specialist, typist	Paper, computer copy machine
Translate some of the handouts	205 days	Agency reps.	Computer with foreign language characters

*Time is measured from the inception of Objective 2, which will begin after completion of Objective 1.

Goal I: To present a culturally appropriate fire safety program to immigrant communities in Districts 5, 8, and 13 in order to decrease the higher-than-average number of fires in these neighborhoods.

Objective I-3: Present a culturally appropriate fire prevention and safety program to the identified immigrant communities.

Tasks	*Time**	*Resources*	
		Personnel	*Equipment*
Train public education personnel in the use of the new program	15 days	Pub. ed. specialist and officer	Newly developed materials
Train agency representatives in the delivery of fire prevention programs	15 days	Pub. ed. specialist and officer	Fire prevention materials and new materials
Schedule presentations inschools, community centers,religious centers, ethnic associations	30 days	Pub. ed. officer, Capt. Herzog	

Tasks	*Time**	*Resources*	
		Personnel	*Equipment*
Present new program in Hispanic and Asian-American communities	210 days	Pub. ed. specialists, agency reps.	Fire prevention materials, new materials

*The total time for accomplishing all three objectives is 18 months.

Goal II: To provide high-quality ALS EMS care whenever called for by residents of the community.

Objective II-1: Formally notify fire fighter/paramedics of the probability of requiring ALS when responding to patients identified in the needs assessment.

Tasks	*Time**	*Resources*	
		Personnel	*Equipment*
Prepare a briefing using the data gathered during the needs assessment	14 days	EMS assistant chief	Copies of needs assessment
Present information concerning track record of ALS required on calls in Districts 5, 8, and 13 to all fire fighter/paramedics	3 days (during standard EMS training each shift)	EMS assistant chief, training officer	Information handouts
Shift captains meet with all fire fighter/ paramedics not present during the training session to inform them of the situation	14 days	Shift captains	Information handouts, attendance rosters

Goal III: Reduce the abnormally high number of EMS calls that require ALS care.

Objective III-1: Inform the leaders of the city and of the immigrant communities about the high proportion of severe accident and illness EMS calls.

Tasks	Time*	Resources	
		Personnel	Equipment
Inform the mayor of the findings:	14 days		
(a) Prepare a brief report using the information from the needs assessment		EMS asst. chief, Chief Sanchez, and maybe EMS chief	
(b) Meet with the mayor mayor and explain the situation			
Present findings to council:			
(a) Work with the mayor to revise report	21 days		
(b) Prepare report for council	30 days		
(c) Send briefing paper to each member of council	30 days	EMS asst. chief, Chief Sanchez	
Present findings to leaders in the minority communities:	60 days		
(a) Identify directors of agencies working with the community		Capt. Herzog	
(b) Contact directors, share information verbally, and ask for names of community leaders		EMS asst. chief	
(c) Send written notification of information to agency directors		EMS asst. chief and Chief Sanchez	
(d) Contact community leaders		EMS asst. chief	
(e) Send follow-up information to community leaders		EMS asst. chief and Chief Sanchez	

Tasks	Time*	Resources	
		Personnel	Equipment
Present findings to appropriate department heads in the city (e.g., building department, planning department, health department, etc.) (a) Identify appropriate departments (b) Prepare a brief written notice (possibly send to council if the mayor approves) (c) Call for informal follow-up of receipt of information	45 days	Chief Sanchez EMS asst. chief	

Note that a number of the tasks are taking place at the same time; they are not necessarily dependent on one another. In addition, the tasks have been broken down into more specific subtasks. A Gantt chart[6] can be used to illustrate this overlap. A Gantt chart is a visual presentation of how tasks relate to one another over time. A sample Gantt chart representing the information action plan forms follows.

Gantt Chart for Goal III, Objective III-1

	Number of Days from Beginning* 7	15	30	45	60
Inform the mayor of the findings:					
(a) Prepare a brief report using the information from the needs assessment	....				
(b) Meet with the mayor and explain the situation		... X			

Gantt Chart for Goal III, Objective III-1 (continued)

	*Number of Days from Beginning** *7*	*15*	*30*	*45*	*60*
Present findings to council:					
(a) Work with the mayor to revise report		....			
(b) Prepare report for council	....	....			
(c) Send briefing paper to each member of council			X		
Present findings to leaders in the minority communities:					
(a) Identify directors of agencies working with the community		....			
(b) Contact directors, share information verbally, and ask for names of community leaders					
(c) Send written notification of information to agency directors			....		....
(d) Contact community leaders				....	
(e) Send follow-up information to community leaders				....	
Present findings to appropriate department heads in the city:					
(a) Identify appropriate departments		....			
(b) Prepare a brief written notice				... X	
(c) Call for informal verification of receipt of information					....

*Note: The ... denotes activity taking place. The X denotes a specific target with a tangible object.

The next action plan form is for planning the achievement of Objective 2 for Goal III. How might you complete it? What other objectives might you design to achieve Goal III?

Goal III: To reduce the abnormally high number of EMS calls that require ALS care.
Objective III-1: Educate about the proper use of the EMS system.

Tasks	*Time*	*Personnel*	*Equipment/ Supplies*

INTEGRATING FIRE CODE INSPECTION REPORTS WITH PRE-INCIDENT PLANNING

Deputy Chief Patterson recently completed an organizational planning course at a local college. As a result of the course, she realizes that the fire department needs to combine the information gathered in the code inspection program with that gathered for pre-incident planning. This would reduce the number of times fire department company or bureau personnel had to visit a given business to gather information. In addition, an integrated information file would reduce the time required to retrieve critical information during an emergency situation.

Deputy Chief Patterson discusses her idea with the chief, who requests that she develop an action plan to accomplish this project. She compiles the following:

Goal: To develop a comprehensive information system that would provide both code enforcement and pre-incident planning data.
Objective 1: Generate a new form that records the data necessary for each activity.
Objective 2: Train company and bureau personnel to use the new form properly.
Objective 3: Incorporate the information in the pre-incident plan used on an emergency scene.

Action Plan

Objective 1: Generate a new form that records the data necessary for each activity.

Task	*Time*	*Personnel*	*Resources*
Schedule meeting of officers responsible for pre-incident planning and code enforcement	1 week	Dep. Chief Patterson, Capt. Code Enforcement, Capt. Pre-incident Planning	None
Identify all information	1 week	As above	None needed
Design data collection form	1 month	As above computer operator for graphics	Personal computer, computer operator

Task	*Time*	*Personnel*	*Resources*
Pilot form review	1 week	Capt. Code Enforcement, Capt. Pre-incident Planning	None
Revise and adopt final form	2 weeks	Dep. Chief Patterson, Capt. Code Enforcement, Capt. Pre-incident Planning, fire chief	Printing

Objective 2: Train company and bureau personnel to use the new form properly.

Task	*Time*	*Personnel*	*Resources*

Objective 3: Incorporate the information in the pre-incident plan used on an emergency scene.

1. Fill in the above action form for Objective 2.
2. Construct an action plan for Objective 3.

ENDNOTES

1. National Fire Academy, "Organizational Theory in Practice" course, *Unit 2: Planning,* written by David H. Hoover, 1991.
2. S. R. Michael, "Organizational Change Techniques: Their Present, Their Future," in *Organizational Dynamics,* II, Summer 1982, pp. 67–80.
3. Henry Mintzberg, "The Structure of 'Unstructured' Decision Processes," in *Administrative Science Quarterly,* June 1976, 21: 246–275. passim
4. Herbert Simon, *Administrative Behavior,* 3rd ed. (New York: Free Press, 1975), 364 pp. passim
5. Adapted from *Unit 2: Planning,* developed by David H. Hoover, for the National Fire Academy course "Organizational Theory in Practice."
6. For further discussion of the use of Gantt charts, see Robert N. Lussier, Supervision: *A Skill Building Approach* (Homewood, IL: Richard D. Irwin, Inc., 1989), pp. 42–45.

REFERENCES

Ascher, William, and W. H. Overholt. 1983. *Strategic Planning and Forecasting: Political Risk and Economic Opportunity.* New York: John Wiley and Sons.

Denhardt, Robert B. 1991. *Public Administration: An Action Orientation.* Belmont, CA: Brooks/Cole Publishing Company.

Graham, Cole Blease, and Steven W. Hays. 1986. *Managing the Public Organization.* Washington, DC: Congressional Quarterly Press, Inc.

Lussier, Robert N. 1989. *Supervision: A Skill Building Approach.* Homewood, IL: Richard D. Irwin, Inc.

Lyden, F. J., and E. G. Miller, eds. 1982. 4th ed. *Public Budgeting: Program Planning and Implementation.* Englewood Cliffs, NJ: Prentice-Hall Publishers.

Michael, S. R. 1982. "Organizational Change Techniques: Their Present, Their Future." *Organizational Dynamics,* II. Summer. pp. 67–80.

Mintzberg, Henry. 1976. "The Structure of 'Unstructured' Decision Processes." *Administrative Science Quarterly.* June. 21: 246–275.

National Fire Academy. 1991. *Organizational Theory and Practice.* "Unit 2: Planning." written by David H. Hoover.

NFPA 1500, *Standard on Fire Department Occupational Safety and Health Program,* 1992 ed. National Fire Protection Association.
Simon, Herbert. 1975. 3rd ed. *Administrative Behavior.* New York: Free Press.

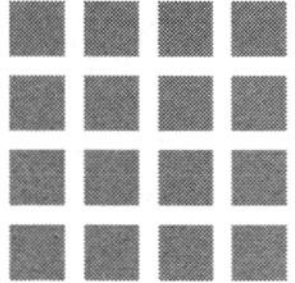

CHAPTER 11

Budgeting

This chapter examines the budget process as well as various types of budgets. The actions to be taken and decisions to be made in allocating funds are presented along with the various formats used to communicate the information to elected officials and to the public. This provides a more comprehensive grounding in the budget process and enables the fire officer to address issues of concern in various departments.

Becoming aware of revenue sources for the governmental entity that funds fire department operations (even if it is an independent fire district) helps fire officers understand the relationship between the amount of funding available to the department and the administration of the overall budget. In addition, understanding the revenue sources enables fire officers to develop a more accurate picture of how much funding could be available for capital and operating expenditures in the future. This is helpful in planning for the future and in making decisions concerning operations.

The preparation and management of a fire department budget is a process that is often misunderstood and undervalued by department administrators. When used properly, budgeting can contribute to increased department efficiency, effectiveness, and accountability. The budget is a visual presentation of the financial status and proposed expenditures of the organization. It is a component of overall planning for the department, since no plans can be put into effect without necessary resources. The budgeting process allocates available funds to the appropriate areas. Thus, budgeting decisions impact the overall operation of the fire department.

The budget reflects changes in the department over time. It must be adjusted to reflect the dynamics of the community. In some instances, rural areas experience growth in population and structures. This creates increased demand for service, including fire protection activities. The budget, in this situation, increases to reflect the expanding services. Growing departments tend to have budgets that increase at a rate higher than the annual cost-of-living increase.

Department downsizing is likely in communities experiencing an economic downturn, and this is reflected in the annual budget. As revenue shrinks, service provision must adjust accordingly. Some programs may be eliminated, or cutbacks may be made in a number of programs. In either case, the budgetary distribution of funds dictates where the cutbacks will be. A community outreach public education program may not be funded, for example. An alternative to this might be to cut all programs by 10 percent. Suppression may have to wait to buy new gloves, fire inspections may have to be rescheduled, and some cuts would still come in public education programs. During times of both growth and downsizing, the budget mirrors the decisions made to expand or shrink services. It is for this reason that the budget, especially when studied over time, is often called the best picture of what a department is doing and how it has changed.

The allocation of funds via the budget process must be tied to the level and types of service offered by the department. Therefore, budget decisions are often equivalent to policy decisions. If a program or service does not receive funding, a de facto policy[1] has been made to cut back or to refuse support to that program or service.

It is important for fire service officers to understand the process of budgeting and to be able to prepare, defend, implement, and oversee the budget. When fire officials fail to understand the complete budget process and focus only on monitoring expenditures, the department often loses its ability to control its own destiny. When public officials outside the department begin to make allocations and assign figures to specific items in the budget without consultation or a thorough understanding of service delivery, they, effectively, determine the future direction of the fire department. Recognizing this, NFPA 1021, *Standard for Fire Officer Professional Qualifications,* identifies knowledge of various types of budgets as a requirement for fire officers. It is imperative that fire service administrators become sophisticated in the budget process.

THE BUDGET CYCLE

The budget cycle is the process through which the budget is made, approved, and expended. It incorporates financial and operational accountability. This process is more formally structured in cities. However, the cycle can also be identified, with less rigor, in smaller districts, villages, rural communities, and townships. The following discussion will first address the formal budget cycle common in cities. Then attention will turn to various budget cycles found in fire departments in less formal government structures.

The Municipal Setting

The budget cycle is a management process involving fire department administrators, municipal managers, and elected public officials. It is a system whereby the issue of fiscal accountability is addressed in terms of "who is accountable to whom?" In a democracy, the general public holds government officials ultimately responsible for resource utilization even though the citizens may not have direct input into the process. Therefore, implementation of the budget cycle process is a joint effort between the fire officer (representing the bureaucracy) and the elected official (representing the general public).

An important third element in this financial process is the expanded role of the professional public administrator or municipal manager. These individuals are usually proficient in handling the technical problems facing the community or society in general. As a result, their role as administrators escalates as a community grows and requires expanded provision of current services, or as the public demands increased services. Conversely, the legislative role has begun to diminish due to the overwhelming technical expertise required to govern the community on a day-to-day basis. The result is that some elected officials will abdicate the more technical functions of the office to a professional manager and, thus, consolidate bureaucratic power under one administrative office.

Less Formal Government Structure

Even in fire districts and townships, the fire chief must construct a budget and present the request to the governing body, which is still accountable to the public. Although there may be no professional public administrators involved, government leaders still play an active role in reviewing the budget requests and questioning the expenditures. This is part of accountability. The discussion of types of governments in Chapter 7, Interacting with Other Public Administrators, provides an overview of who might be involved in the budget process in different types of governmental organizational structures. Realize that even if the process is informal, the steps discussed in this section are followed to some extent every time a budget is put together.

Steps in the Budget Cycle

In order for the budget cycle system to function effectively, four elements must be addressed in a logical sequence. They include: (1) budget preparation, (2) adoption of the budget, (3) execution of the budget, and (4) audit and evaluation. Although each is a separate phase, they often tend to overlap since government is an ongoing process that does not operate within segmented planning or fiscal boundaries. For example, budget preparation for the next fiscal year may be initiated before the evaluation phase of the current budget period is complete.

Budget preparation is the first phase in the process. This activity involves the careful assessment of all fire department expenditures for programs that have been funded during the current fiscal year. Care should be taken to ensure that the same criteria are used to assess *all* programs. Fire executives must be prudent in their evaluation of each program and not become entrapped by either tradition or popular trends.

One way to avoid this problem is to establish an evaluation form that is structured to ask the same questions of each expenditure or program and then categorize the answers in a usable fashion. For example:

1. *Programs to be retained.* Is the program doing what it was intended to do?
2. *Programs to be retained with modifications.* Will additional resources be required? Do the modifications require establishing new objectives? (For example, the department may want to change the nature of a public education program that was using only fully certified fire fighters to one that involves the use of fire cadets. This program is being modified, not expanded or diminished.)
3. *Programs to be increased.* What new resources will be necessary in terms of staffing and funding?
4. *Programs to be de-emphasized.* Is it necessary to re-establish the objectives? What percentage of the resources allocated to the program can be deleted from the next budget?
5. *Programs to be deleted.* Is it possible to substitute the resources allocated to this program to a new or existing program?
6. *New programs to be instituted.* What level of resources will be required for the initial start-up and ongoing activities? Can the program goals and objectives be tied to resource allocation?

This evaluation process should include input from all officers and unit managers responsible for the administration of a division, shift, bureau, or special program. It is also important to include information from the fire department's strategic plan. (See Chapter 10, Planning.)

To measure the effectiveness of a program accurately, the *results* of the program and *not* just the *activities* conducted must be measured. For example, the statement "during the previous year, the department's fire prevention bureau staff performed 3,561 business and mercantile inspections" is an accurate measure of bureau activities; however, it does not provide information pertaining to the results achieved by these efforts and what benefits the department will realize this year and in succeeding years as a result of these inspections. Results can be measured and presented a number of ways in order to demonstrate the impact of a program. They can highlight a reduction in the number of fires that occurred in commercial occupancies or the number of fire or building code violations that were corrected during the year. For example: "75 percent of all business and 90 percent of all mercantile violations cited were corrected within 120 days." This demonstrates accomplishment rather than activity.

The evaluation of current programs can be incorporated into the planning functions of the department. As members of the department review each program, the *plans* for what to do with the program in the coming year(s) become part of the department plan. At the same time, however, this evaluation yields decisions about whether a program should continue to receive funding. Basically, the fire officers, and perhaps other members of the department, are evaluating whether current expenditures of department funds are going to the most effective aspects of the operation.

Even in smaller departments, this evaluation of programs and of monetary expenditures must take place. For example, a small rural fire department has been planning to purchase a new portable pump for the grass fire truck. The number of grass fires in the district, although high, has not increased over the past 3 years. However, the opening of a new interchange bypass has resulted in a significant increase in the number of motor vehicle accidents. The nearest hydraulic rescue tool is in a department 50 miles away. The department may want to shift funding from the grass truck to the hydraulic rescue tool. This happens when the department *evaluates* the planned budget expenditures. Even if this is an informal process where the members of the department sit down, look at what they had planned to purchase, talk about the runs they have been having, and take a vote to change what is being purchased, they are evaluating programs in order to prepare the budget.

Once the evaluation task is complete, department expenditures can be matched to specific programs and projections can be made. The end product should be a department budget that accurately projects the resources needed to operate during the next fiscal year along with a detailed plan of action for funding allocations.

The budget should be directly tied to the plan that has been adopted by the department and the governing entity, as appropriate. The elements of the budget should correspond to the goals and objectives adopted by the department in its planning process. Once this is done, the budget demonstrates the cost of fulfilling the approved elements of the plan and details where the money will be spent.

This budget document may take a variety of forms, as will be discussed later. What is crucial at this point in the process is that it reflects the decisions made concerning the operation of the department.

Adoption or approval is the next step. This phase is the legislative process whereby the proposed budget and plan for resource allocation is approved. The approval process usually requires a formal presentation of the proposed budget to the council, township board of trustees, district governing board, or other legislative entity. Detailed information, including a description of the proposed expenditures, a comparison of the previous year's actual expenditures, and an estimate for the current year, should accompany the budget. Once the budget is presented to the legislative body by the municipal manager or the fire chief, it then becomes an issue for legislative debate. Because a balanced budget is a legal requirement for local and state governments, the manager usually recommends a budget that is balanced. Should the legislature desire to increase expenditures in one fire department program, they will be forced to reduce the expenditures in another in order to maintain a balanced budget. This action is generally held to a minimum because large portions of the budget address the technical aspects of fire department operations, which are beyond most legislators' level of expertise. Sometimes, however, the legislative body will decide to cut spending in one department in order to provide more funding to another. The fire department may gain or lose in this situation.

Depending on the organizational structure of the community government, it may be necessary for the fire chief to present the budget to intermediary administrators for approval prior to presentation to the legislative body. Some cities, villages, counties, and townships utilize positions such as safety director, deputy mayor in charge of public safety services, or chief administrator. In these instances, the budget must first be presented to this office for consideration. The fire department budget will then be incorporated with other department budgets. Thus, the legislative body that holds the authority to allocate funding views the fire department budget as only one part of the entire budget.

In fire districts, the fire chief is usually called on to present the budget to the governing board. In these instances, the chief should take the same steps of preparing information to accompany budget requests. In smaller communities with volunteer or combination fire departments, the fire department sometimes relies on two sources of funding: (1) the local legal government entity (township trustees, village council, etc.) and (2) the fire department association or auxiliary, which conducts various fundraising activities for additional equipment or desired items. In these instances, the fire chief or group of fire officers must still request allocation of funding, but from two sources. The approval of funding from the association may well take the form of a membership vote. This may be done in a very informal atmosphere; however, it is still a part of the budget cycle and a form of budget approval. The association basically allows for the expenditure of funds on certain approved budgetary items and can choose to allocate its funding in any way it desires.

Execution or spending is the third phase of the process. Once the budget has been approved, the fire department receives its operating budget in the form of an appropriation. If the legislature or intermediary administrator has revised portions of the budget, the department must adjust the original budget to meet the requirements set forth. Funding is allocated to precise accounts as detailed in the approved budget. Specific expenditures are then recorded against the appropriate accounts in order to provide an accurate ongoing balance. If the fire administrator wishes to modify the approved budget, there are certain methods by which this can be accomplished. Fund transfers and budget modifications are discussed in more detail on page 373.

Although the amount of funds available and the degree of discretion in spending may vary from department to department, the procedure of tracking expenditures versus allocated funds is the same. Departments must have a system of monitoring expenditures versus available budget amounts to ensure that the department does not go into debt and that the money goes to purchase services and equipment that had been approved when the money was allocated to the department.

Audit or review is the final phase. The types of audits include: fiscal, operational, program, performance, and management.

The **fiscal or financial audit** is conducted to determine whether monies appropriated by the approval phase were expended legally. Independent auditors may be hired to conduct the audit, or it may be performed by internal auditors. Generally, even when internal auditors are employed, there is a periodic review by an external auditor or certified public accountant (CPA) as well. The process does not necessarily involve the examination of every transaction, but instead will use a statistical sample selected by category of transaction. In departments with small budgets and few categories, this audit may take the form of matching purchase orders and receipts with deductions for the amount of funding in the department, at the same time checking to see that what was purchased had been authorized.

Operational and management audits focus on management policies of the fire department and the administration of the municipality. A management audit will usually evaluate the performance of managers and contain recommendations for changes in policy or improvement. Operational audits will examine specific programs or tasks in terms of what objectives are implemented to accomplish the desired outcome. This is similar to a program review or evaluation. Both of these audits are often incorporated into internal needs assessments and evaluations. (See Chapter 9, Needs Assessment.)

The **performance audit** is similar to the exercise conducted in phase one of the budget cycle. It addresses the issue of program performance in terms of past accomplishments, current budget allocations, and future expectations. "What is the program accomplishing?" is the question most often applied to this type of evaluation.

Audits are conducted to be certain that allocated funds have been dispersed and accounted for properly. In addition, audits of management, operations, and programs ensure that department and municipal policies and activities address the mission of the agency and are directed toward achieving the goals set forth by fire administrators.

This phase is often incorporated into the planning process in order to match the funding with the programs or activities deemed to be most appropriate for meeting the needs of the community.

Even small communities and volunteer fire departments should have some form of audit. Informal audits can be a simple review of expenditures compared with allocations that is presented to the community and members of the department at the end of each fiscal year. If nothing more, the audit should function as a review of what has been done with public funds at the end of a year.

OPERATING AND CAPITAL BUDGETS

Operating and capital budgets deal with two different types of expenditures in a fire department. As such, these are separate documents, and although the allocation in one may affect the costs in the other, they are prepared and executed separately.

Operating budgets are annual documents detailing allowable and expected disbursement for the fiscal year.[2] Expenditures from this budget pay for the daily expenses of providing all fire department services to the community. Items included in the operating budget are: (1) payroll, (2) fringe benefits, (3) expendable or small equipment, (4) equipment maintenance, (5) station supplies, and (6) operating overhead, which may include a contribution to the operation of a shared dispatch center or fees for use of the central computer system. The operating budget is formulated, approved, and executed in a 1-year cycle. Carryover of excess or unused funds to the next fiscal year does occur in some instances. For example, when the timing of personnel replacement results in a vacant position for 4 months, that portion of the salary remains in the budget and becomes surplus at the end of the year. Another example might occur when hand tools are bought through a cooperative purchasing agreement that results in a 15 percent savings. In both instances, the surplus would be carried over into the next fiscal budget. However, since operating budgets attempt to match allocation of funds with needed expenses, a large amount of surplus is generally considered a sign of inefficient budgeting.

Capital budgets include expenditures for items, such as fire apparatus, that will be used over a period of several years. Capital items have a higher cost and are paid for over a period of time greater than 1 year. As Robert Denhardt points out, the primary reason to separate consideration of capital items is that the benefits of these items are spread over future generations and, therefore, the burden of repaying the money borrowed can be reasonably shared.[3] The high cost of most capital expenditures often necessitates either issuing a bond or accumulating funds over a long period of time. Depending on the financial status of the local community and the quality of financial resource planning that has taken place, a community may need to authorize debt to purchase capital items.

Capital expenditures can result in the acquisition of or addition to fixed assets, including construction of buildings or improvements, the purchase or refurbishing of apparatus, land acquisition, and large-scale equipment purchases. They also include replacement and major alterations to fixed facilities and structures. Capital expenditures do not include minor equipment of insubstantial monetary value or those that have a life expectancy of less than 5 years.

The decision as to how to allocate capital expenditures usually involves long-range planning and decision making rather than annual, short-term allocations. It must not be a one-time attempt to solve department apparatus or space needs. Instead, capital budgeting is an ongoing process covering, on average, a 5-year cycle. As the first year of the 5-year cycle is completed, a new one is added to the end and plans are updated. This enables the administrators to adjust expenditure projections to match changes in service needs and financial capacity.

Approval of capital expenditures must be based on the current and long-term needs of the community. The first step is to identify what capital expenditures will be needed over the next 5 years. This identification can be accomplished in a variety of ways, indicating a complete community needs assessment, individual internal/department needs assessments, citizens capital funding task force findings or counsel, or other government-elected decree. Once this is done, the cost for each project is estimated and any restrictions attached to revenue sources are identified. For example, some fire levies specifically state that the funds are to be used for operating expenses or personnel. Thus, these cannot be used for building a station, which would be a capital expenditure. Other levies specify equipment purchase or capital improvements. These restrictions must be clearly presented when decisions for capital project funding are made.

The second step is to assess the degree to which the capital items will improve service to the community. This is generally done through a potential benefits analysis, which identifies how service will be improved and which citizens will benefit from improved services. Once this is accomplished, the third step is to rank all capital projects in priority based on the relative needs of the community. Capital programs are identified as being: (1) urgent, (2) essential, (3) necessary, (4) desirable, (5) acceptable, and (6) deferrable. Allocation of funds is made on the basis of this priority until revenue is exhausted. Capital programs that are deferred may enter the cycle the next year.

PRIORITY RATING MATRIX

The priority rating matrix system can be used to make decisions concerning capital expenditures to be made in any given budget cycle. There are many systems or techniques to assist in making budget allocation decisions. Most are based on the same principles of priority setting incorporated in this matrix system. The priority rating matrix considers a variety of factors in both a visual and a logical manner. As such, it has proven to be effective in providing information for budgeting decisions. This technique helps fire officers arrange the information analytically by categorizing the relative benefits of various purchases.

The priority rating matrix on page 364 has been adapted to illustrate capital budget decisions faced by fire departments.[4] This system has been used by a variety of municipalities and departments. It incorporates community priorities in a prominent position in the decision-making process and yet balances them with the requirements of legislation.

Fire departments can use this matrix to begin to identify and defend the relative demands for various capital expenditures. The matrix also provides a forecast of probable future capital expenditure needs. This can then be incorporated into the department's planning process.

Funding for capital improvements must be identified through a system of prioritization based on the relative importance of the needs of the community.

The priority rating matrix is set up by listing the department functions in the first column. These functions are arranged in order of the community service priority rating (column 2). This departmental ranking identifies which functions the fire department rates as most important. They are grouped according to major service function, with the specific

Priority Rating Matrix for a Fire Department

Degree of Urgency:		*State/ Federal Mandate*	*Council Commissioner Mandate*	*Community Hazard*	*Regional Cooperation*	*Standard of Service*	*Increase in Fire Safety*
		Degree of Urgency Categories					
Department Function:	*Community Service Priority Rating*	*HIGH 1*	*2*	*3*	*4*	*5*	*LOW 6*
Prevention							
Inspection							
Vehicles	1					@	
Public Education							
Equipment	2		@			@	
Suppression							
Apparatus	3	@					
Equipment	4	@				@	
Stations	5				@	@	
Personnel							
Equipment	6			@			
Hazardous Materials							
Apparatus	7		@				
Equipment	8	@					
Personnel							
Equipment	9		@				
EMS							
Vehicles	10		@				
Stations	11				@		
Equipment	12					@	

capital items then listed. For example, fire prevention has been identified as the single most important program of the department in the sample matrix. Within this program there are two primary activities: inspections and public education. Within each of these there is a need for capital expenditures for equipment. Of these, vehicles is rated 1 and equipment (for public education) is rated 2. This continues until all capital equipment requests are listed in order.

For the community used in the example, EMS is listed last. It is the lowest capital funding priority in the community this year. This does not mean that EMS is a low-priority service. Remember that this is the capital, not the operating, budget that is under discussion. The low ranking demonstrates that EMS is not seen as a service that needs capital expenditures at this time. This could be because the community purchased three new advanced life support squads last year. Thus, the service does not need more *capital* expenditures this year—at least not as much as the other services do.

The top of the matrix contains six categories that identify the reasons for purchasing equipment to provide the associated service. If providing the service is a state or federal mandate, it is designated as having the highest degree of urgency, because it is law. The second highest degree of urgency is that required by council or commissioner [or other local elected official(s)]. The third is that necessary to address a community hazard. The fourth level of urgency deals with improving or maintaining regional cooperation. The fifth category is that of maintaining a standard of service, while the sixth and least urgent degree of urgency is that of acquisitions that will increase the existing level of fire safety. An increase in level of service is generally seen as a preferred improvement rather than necessary. Thus, the lower the number, the higher the priority. This is indicated by the heading under "Degree of Urgency" at the top of the matrix. The continuum runs from high, equal to 1, to low, equal to 6, in urgency.

A community can rearrange the order of these categories by defining the degrees of urgency according to their service priorities. However, the first three categories should never be changed. It is first and foremost the responsibility of every local fire department to fulfill federal and state law. Following that, it is mandatory to follow local mandates. The third most important aspect of service is to address known community hazards. After these three degrees of urgency categories, the importance of regional cooperation, maintaining standard of service, and increasing fire safety can vary from one community to another.

Steps to Constructing the Priority Rating Matrix

1. The department ranks each of its programs in order of importance. For example:
   (a) Prevention
   (b) Suppression
   (c) Hazardous materials
   (d) EMS
2. Within each program it rank-orders the capital expenditures necessary for that year.
   (a) Prevention
      (i) Vehicles
      (ii) Equipment for Public Education
   (b) Suppression
      (i) Apparatus
      (ii) Equipment
      (iii) Station
      (iv) Personnel equipment
      etc.
3. These are placed along the left margin of the matrix, in the first column.
4. Once put in order, each is numbered according to the community service priority rating (see column 2):
   Prevention—vehicles 1
   Prevention—public education 2
   Suppression—apparatus 3
   Suppression—equipment 4
   to end >
   EMS—stations 11
   EMS—equipment 12
5. The degree of urgency category is then identified and marked with an agreed-upon symbol for each individual purchase (the

above matrix uses the "@"). For example, the purchase of the prevention inspection vehicle is necessary in order to maintain the current standard of service (category 5). The public education equipment is required by a council mandate that each public school and senior citizen center be presented with a fire safety program at least once each year. The hazardous materials equipment acquisition is required under the federal and state mandates and is, therefore, identified as necessary under urgency category 1.

It is possible to have more than one justification for purchasing items. Suppression equipment is needed in order to continue to provide the current standard of service (category 5). It is also a state compliance issue in states that have adopted NFPA 1500, *Standard on Fire Department Occupational Safety and Health Program,* and is identified by the "@" under urgency category 1. For example, a community may need to increase the number of SCBAs in order to have enough units for each fire fighter exposed to hazardous atmospheres. This would be justified under both category 5 and category 1.[5]

6. Once the prioritization is complete, the capital improvement priority is calculated. This is done by multiplying the community service priority rating by the number in the degree of urgency category. *When two or more categories are identified, always use the one with the greatest degree of urgency—the lowest number.*

   Thus, the computations for the example would produce the results shown on page 369. This capital improvement score is then used to rank-order capital improvement expenditures in terms of funding priority.

7. Items are rank-ordered from lowest capital improvement score to highest. When two items have the same score, the one having the higher community priority is ranked above the other. For example, prevention public education equipment is ranked 2, ahead of suppression equipment, which is ranked 3. The lowest scores are the most important items and will be purchased first. The higher-scoring items will be purchased only

Capital Improvement Score Computation

Program Item to Be Purchased	*Community Service Priority Ranking*		*Degree of Urgency Category*		*Capital Improvement Score*
Prevention					
Inspections					
Vehicles	1	x	5	=	5
Public Education					
Equipment	2	x	2	=	4*
Suppression					
Apparatus	3	x	1	=	3
Equipment	4	x	1	=	4*
Stations	5	x	4	=	20*
Personnel Equipment	6	x	3	=	18
Hazardous Materials					
Apparatus	7	x	2	=	14
Equipment	8	x	1	=	8
Personnel Equipment	9	x	1	=	9
EMS					
Vehicles	10	x	2	=	20*
Stations	11	x	4	=	44
Equipment	12	x	5	=	60

*Note that there are two items with capital improvement scores of 4 and two with scores of 20. Their order in the next ranking is determined by community service priority, as explained in step 7.

if there is enough funding available. Thus, the resulting capital improvement funding priority ranking differs somewhat from the community service priority ranking. This is due to the influence of legislative mandates. See the example on page 370.

8. Capital funds are then distributed to purchase the identified items until all available funds have been allocated. The remaining items are relegated to future capital budget purchases. Looking at hypothetical costs of the first five items illustrates this procedure.

Capital Improvement Funding Priority Ranking

Funding Rank	*Program*	*Community Service Priority Rank*
1	Suppression apparatus	3
2	Public education equipment	2
3	Suppression equipment	4
4	Fire prevention vehicles	1
5	Hazardous materials equipment	8
6	Hazardous materials personnel equipment	9
7	Hazardous materials apparatus	7
8	Suppression personnel equipment	6
9	Suppression stations	5
10	EMS vehicles	10
11	EMS stations	11
12	EMS equipment	12

1	Suppression apparatus	$675,000.00
2	Public education equipment	75,000.00
3	Suppression equipment	240,000.00
4	Fire prevention vehicles (2)	35,000.00
5	Hazardous materials equipment	8,000.00

If the department has been allocated $1 million in capital funds, it can purchase the suppression apparatus, the public education equipment, and the suppression equipment. These purchases total $990,000, leaving $10,000. This is not enough to purchase even one of the fire prevention vehicles. It will, however, cover the cost of the hazardous materials equipment. Therefore, items ranked 1, 2, 3, and 5 are purchased during this budget cycle. Any remaining funds, in this case $2,000, can either be carried over to the next annual budget or expended on items identified on the approved list. In this case, some of the suppression personnel equipment (ranked sixth) could be purchased, but only items equal to the remaining $2,000.

This matrix is used to incorporate department priorities with existing degrees of funding urgency. There is never enough capital funding to cover all items a department needs or wants. Ranking capital expenditures in order of funding priority enables the fire administrator to prepare a more realistic long-range plan for the department. With the many competing responsibilities facing the fire service, it is important to make objective decisions when allocating scarce resources. Meeting legislated obligations and maintaining the safety of the community are paramount. The priority rating matrix is a tool that fire service administrators can use to maximize the use of existing capital funds and to prepare for the future.

BUDGET TYPES

There are various types of budgets in use around the nation. The most common—line-item or executive, program, performance, and zero-based budgeting (ZBB)—will be discussed here. In addition, the planning-programming-budgeting system (PPBS) will be presented following the section on program budgeting.

Each budget type has positive and negative aspects. Since each was developed to serve a somewhat different primary purpose, one may serve a particular community better than another. Perhaps a combination of budget types will be appropriate: a community may incorporate some principles of a line-item budget with a program budget to exert more control over expenditures. Thus, the question, Which budget is best? needs to be rephrased as, Which budget is best for the community?

Most fire departments must use whatever type of budget the community governing structure or funding entity mandates. However, it is good to understand the options available as well as the principles and technical aspects of the one in use. This is one area where smaller fire departments or fire districts may have more autonomy and be able to decide or influence what type of budget will be used. The principles behind each of the budgets described herein explain why they are used. Even though fire officers may not be able to actually use some of the budget techniques presented, they may be able to apply some of the principles behind the techniques as they make budget allocation and expenditure decisions. This practice would improve the efficiency as well as the effectiveness of fire department operations.

Line-Item Budgets

The line-item or executive budget was initially established in order to provide control over the expenditures of public officials. Until the early 1900s, municipal administrators had total control over the public purse. They could spend money without justification and often did not have to tell anyone what was being purchased. Abuses of this system led to reforms designed to control the actions and expenditures of public officials. The line-item budget was developed during this era.

In a line-item budget, each type of expenditure is allocated a separate line in the budget, which is simply a list of things to be purchased. The adoption of the budget gives the administrators authorization to spend the funds *for the item designated in the line.* In this way, authorization of $200 for office supplies can be used to purchase paper, paper clips, pencils, pens, etc., in any combination. However, none of this money can be used to purchase hose, gloves, or hand tools, which would be contained under another line item: loose fire fighting equipment.

Line-item budgets can impose strict control or loose, general guidance, depending on the detail involved and the procedures for transferring funds among different categories. The nature of the line-item budget determines the amount of purchasing discretion allowed to the administrator. Generally, as a budget becomes more detailed, the level of control increases and the level of administrative discretion decreases. The following budget segments illustrate this phenomenon.

Line Item Budgets

Example 1:	
Supplies	
Office	$500.00
Fire suppression	$1,500.00
EMS	$2,000.00
Total	**$4,000.00**
Example 2:	
Supplies	
Copy toner and paper	$350.00

Desk supplies	$150.00
Personal protective equipment	$500.00
Hand-held tools	$750.00
Hose clamp	$250.00
Portable stretcher	$1,250.00
Portable oxygen supply	$750.00
Total	**$4,000.00**

The first example gives the administrator much more flexibility in purchasing items for the department. There is very little discretion left to the administrator in the second example. Almost every expenditure is carefully controlled by the narrow definitions of the line items.

Transfer of Line-Item Funds

The policy for transferring funds from one line item to another is the other characteristic of line-item budgets that determines the level of control exerted over fire department administrators. The ability of fire service administrators to transfer funds among different line items in the budget varies from one jurisdiction to another. Differences are generally defined by three factors:

1. *The amount of funds that can be transferred from one line-item category to another.* This amount is either a percentage of the total budget, a percentage of a given line item, or a fixed dollar amount. Local policies mandate which method is used to determine the amount of funds that can be transferred once the budget has been approved. For example, the fire chief in one community may be allowed to transfer up to 10 percent of the total department allocation while in another community only a total of $1,500 can be transferred in any given fiscal year. This is another indication of the degree of control policy makers hold over the local department administrator.

   In smaller departments the chief may or may not have more discretion. When much of the funding is raised from outside sources such as gifts from the fire association or state grants, what the money can be spent on is directly determined by the purpose designated in its allocation. For example, a state grant for training for hazardous materials incidents to comply with SARA Title III cannot be used to buy gloves or to train paramedics. On the other hand, because of the relatively small size of the budget and the informal governing structure, some fire chiefs are given almost total control to allocate funds in any way they wish.

2. *The procedures to be followed in transferring funds.*[15] The procedures to be followed dictate who is involved in the process and what type of documentation is required for the transfer. The number of people involved in the process, and what position they hold, influences the ease of transfer and illustrates the level of external control being placed on the fire administrator. For example, a chief who can sign a transfer form that is then sent to the finance department and processed has much greater discretionary budget authority than one who has to secure approval or permission from the finance director and elected official, such as a council person. The documentation necessary to transfer funds may range from a single one-page form requesting transfer to a set of complicated forms that require a full narrative explanation of why the funds will be transferred.
3. *Regulations governing the source of the funds, e.g., state grants or dedicated fire levies.* Some state grants limit the amount of funds that can be transferred without state approval. These restrictions may apply to grants for training only, for acquisition of specific types of apparatus, and for communication equipment. Also, certain fire tax levies are dedicated for specific items and cannot be spent on other needs. Fire departments will request voter approval for specific equipment acquisitions, such as EMS or hazardous materials equipment. Although the funds may appear in the line-item budget, the law prevents using these funds for anything other than that stated in the levy.

Thus, the line-item budget indicates not only the amount of funding available, but also the amount of discretion the fire service administrator has. It is apparent that the purpose of the budget determines its design.

Some aspects of the line-item budget are found in almost every budget type. Personnel and nonpersonnel line items are listed by program, for example. Thus, people often confuse the type of budget used simply because items are listed or presented in "lines."

Program Budgets

The program budget is organized according to the programs operated by the department. This budget illustrates what the department is *doing.* The allocation for each program is presented separately, showing the amount of resources allocated to each.

Thus, each program has its own budget and it is possible to see the relative costs of the programs and where resources are allocated. A program is an arrangement of department resources in order to achieve a given objective. These objectives are generally based on preventing or solving an identified problem. Programs usually extend beyond one fiscal year; they tend to be ongoing activities.

Program budgeting is one way to more closely match long-range planning and decision-making activities with the daily operation of the department. As personnel perform the program activities, they are reminded of the objectives and department goals that they are trying to achieve. Elected officials are reminded of why they are allocating funds to a given program. Program budgeting ties more directly to the long-range planning, as it evaluates the programs adopted in the plan and the goals and objectives these plans are designed to achieve.

Program budgets have been adopted or adapted by many communities. Although the strict guidelines of program budgeting are not always followed, many communities require that the primary expenditure categories be broken out among the different programs in place. This presents the activities of the community in a more positive and understandable format. In addition, tracking the budget from one year to the next by examining the changes in the different programs informs administrators when costs of operation are increasing at a faster rate than those of other programs. This may signal the need for a performance review. Likewise, a particularly efficient program may be identified and reviewed to determine whether it demonstrates more desirable methods of operation that can be applied to other programs.

In summation, because the program budget is fairly easy to prepare and understand, it is popular with both administrators and elected officials. Policy analysts and citizens also like the program budget because they can see where the money is being allocated, i.e., which policies are being supported by the government.

Performance Budgets

Performance budgets are similar to program budgets. Performance budgets focus on activities rather than comprehensive programs. A performance budget would tell the public official how many inspections could be performed for a certain amount of funding, while the program budget would present the cost of various prevention activities. (Sometimes, however, program budgets get lost in "bean counting" and lose sight of *why* activities are being performed.)

Much of the terminology is the same and, as program budgets have become more popular, the use of performance budgets is rare today. Increased use of planning has developed greater awareness of the goals and objectives of specific activities. Thus, attention focuses not only on per unit or activity cost, but on the *results* of the combined activities as well. This, therefore, can be compared with the goals and objectives adopted in the planning process.

The **Planning, Program, and Budgeting System (PPBS)** is an attempt to integrate the planning and budgeting processes in a systematic manner based on specific programs enacted to achieve the goals and objectives established in the planning process. Allocations are made based on the priorities adopted during planning. Too often, plans call for accomplishing one activity while resources are not provided to that program. PPBS establishes a system of tying resource allocation directly to the goals and objectives.

Zero-Based Budgeting (ZBB)

Zero-based budgeting attempts to focus decisions on what impact different levels of funding will have on the delivery of services. It requires a presentation of the impact of budget cuts and increases in terms of the type and level of service. In this way it directs the decisions on the results of allocations rather than on the amount of allocation.

Each program is required to prepare three budgets, one for continuing operations at the current level of funding, one with increased funding, and one with a cut in funding. The zero in the name of this budget format refers to the fact that the original, "pure" form required program administrators to identify the impact of funding the program at zero level ($00.00). In practice, however, the two alternative budgets are generally based on a 10 to 20 percent increase and decrease.

Under ZBB, each program has to relate its activities to the most current priority list established by the legislative body. This requirement applies to existing as well as new programs and is designed to encourage transfer of funds from programs operating inefficiently to those that would use the resources better.

This format is not commonly used due to the amount of time involved. In addition, administrators often feel vulnerable when they have to document what level of service they could provide with a decrease in funding. There is always the possibility that the elected officials may decide that the decrease provides an acceptable level of service for the community. However, in times of financial crisis and cutback management, ZBB offers a way of identifying which services will be affected and how. It also forces elected officials to identify which programs will be affected rather than leaving this decision to the department administrators.

In recent years, program budgeting has become the most popular format. It presents an overview of the activities performed by the department and their relative costs. In this way, as priorities change, resources can be shifted from one program to another. At the same time, however, it is not so detailed as to be unwieldy and complicated.

Program budgets can be tied to long-range department or community plans by addressing the objectives adopted. In addition, an evaluation of the success of activities can be performed on a program-by-program basis. Questions of efficiency and effectiveness are thus focused on programs rather than on total department performance. The department can still take the responsibility for improving programs.

Each type of budget has its strengths and weaknesses. The key goals of good budgeting are the same:

1. To get the most for the money
2. To know why money is being spent the way it is
3. To be able to identify where the money went (tracking and audit)
4. To be able to justify why the money was spent (accountability)
5. To be able to identify the expenditures in terms of service to the community

Understanding different budget systems, especially in light of the planning process, will help fire officers understand why decisions are made and what some of the major driving and restraining forces facing department operations are.

Examples of Different Budget Formats

The following examples of different budget formats illustrate how equal amounts of funding would be presented using the different formats discussed. The first budget, the line-item budget, shows only the totals of all personnel. As a result, it is not possible to identify how many personnel, or how much money, is allocated to fire prevention. The program budget format demonstrates how the $130,175.00 allocated to fire prevention is spent. Note that this would mean that $1,931,825.00 is spent on programs *other than* fire prevention. Each of these other programs would also have a program budget of its own. Together they would account for the total department allocation.

The zero-based budget format reduces the budget used in the example even further, demonstrating only how the fire inspection program component of the fire prevention program is budgeted using ZBB. These examples clearly demonstrate the amount of detail and work required for each budget type. The amount of time associated with the preparation of each becomes obvious.

The following questions should be considered when comparing the three formats:

1. Which budget tells more about the different things the department does?
2. Which budget demonstrates more discretionary control on the part of the fire administrator?
3. Which budget would be most helpful in planning future activities?
4. Which budget format do you currently use?
5. Which would you like to use? Why?

LINE-ITEM BUDGET FORMAT

Fire Department	
Personnel	
Administrative	130,000.00
Clerical	38,000.00
Dispatch	74,000.00
Fire fighters	1,021,000.00
Inspectors	58,000.00
Public education	21,000.00
	1,342,000.00
Fringe @ 30%	402,000.00
Total Personnel	**$1,744,000.00**
Nonpersonnel Expenditures	
Apparatus maintenance	50,000.00
Tools and equipment	150,000.00
Personal protective equipment	75,000.00
Station maintenance	25,000.00
	300,000.00
Nondiscretionary	
Utilities	18,000.00
Total Nonpersonnel	**$43,000.00**
Total	**$2,062,000.00**

PROGRAM BUDGET FORMAT

FIRE PREVENTION

Fire Inspection Program	
Personnel	
Inspectors 2@ $29,000.00	58,000.00
Clerical 1/2 time @ $18,000.00	9,000.00
	67,000.00
Fringe @ 30%	20,100.00
Total Personnel	**$87,100.00**
Nonpersonnel	
Fuel for inspector vehicles	1,200.00
Maintenance of inspector vehicles	400.00
Office supplies	200.00
Total Nonpersonnel	**$1,800.00**
Program Total	**$88,900.00**

****

Public Education Program	
Personnel	
Public education specialist	21,000.00
Clerical 1/2 time @ $18,000.00	9,000.00
	30,000.00
Fringe @ 30%	9,000.00
Total Personnel	**$39,000.00**
Nonpersonnel	
Fuel for vehicle	300.00
Maintenance of vehicle	70.00
Office supplies	100.00
Printed handouts	1,500.00
Displays	500.00
Presentation materials	300.00
Training program fees	65.00
Total Nonpersonnel	**$2,835.00**
Program Total	**$41,835.00**
Fire Prevention Division Total	**$130,175.00**

ZERO-BASED BUDGET FORMAT

FIRE INSPECTION PROGRAM—Current Level of Service

Personnel	
Inspectors 2 @ $29,000.00	58,000.00
Clerical 1/2 time @ $18,000.00	9,000.00
	67,000.00
Fringe @ 30%	20,100.00
Total Personnel	**$87,100.00**
Nonpersonnel	
Fuel for inspector vehicles	1,200.00
Maintenance of inspector vehicles	400.00
Office supplies	200.00
Total Nonpersonnel	**$1,800.00**
Program Total	**$88,900.00**

Accomplishments per Year

Percentage of Inspections Completed by Occupancy Type

Place of public assembly	90%
Mercantile	75%
Education	100%
Industry	80%

****

FIRE INSPECTION PROGRAM—Increase of Funding by 20%

Personnel	
Inspectors 2 @ $29,000.00	58,000.00
Clerical 1/2 time @ $18,000.00	9,000.00
	67,000.00
Fringe @ 30%	20,100.00
Part-time inspector—60% time (no fringe)	17,230.00
Total Personnel	**$104,330.00**
Nonpersonnel	
Fuel for inspector vehicles	1,500.00
Maintenance of inspector vehicles	600.00
Office supplies	250.00
Total Nonpersonnel	**$2,350.00**
Program Total	**$106,680.00**

Accomplishments per Year	
Percentage of Inspections Completed by Occupancy Type	
Place of public assembly	95%
Mercantile	90%
Education	100%
Industry	85%

****

FIRE INSPECTION PROGRAM—Decrease of Funding by 20%

Personnel	
Inspector (1)	29,000.00
Clerical 1/2 time @ $18,000.00	9,000.00
	38,000.00
Fringe @ 30%	11,400.00
Part-time inspector—1 @ 70% (no fringe)	19,920.00
Total Personnel	**$69,320.00**
Nonpersonnel	
Fuel for inspector vehicles	1,200.00
Maintenance of inspector vehicles	400.00
Office supplies	200.00
Total Nonpersonnel	**$1,800.00**
Program Total	**$71,120.00**

Accomplishments per Year	
Percentage of Inspections Completed by Occupancy Type	
Place of public assembly	75%
Mercantile	50%
Education	100%*
Industry	70%

*Note: 100% inspection mandated by law.

SOURCES OF REVENUE

Many fire officers are never involved in raising money for the department. Others find that fundraising is a constant and important part of their job. In either case, it is important to understand how money can be raised. Many fire departments, especially volunteer and combination departments, are funded by special levies. The passage of such levies, their adequacy in terms of providing enough funding, and their continuation are often mysteries to fire officers. They fail to understand how funding levels are tied to income levels and property values, which may fall as well as rise. In addition, support for fire department levies will be influenced by other taxes imposed on community residents. When resistance to levies is met, it does not mean residents do not support the fire department, it simply means that someone else got the money first and the residents do not feel they can afford more taxes to support the department.

This section deals with the funding sources to which fire departments have access individually or as members of large government organizations. While fire officers may not be directly involved in selecting or attempting to implement these sources of revenues, understanding how they operate and what their strengths and weaknesses are helps explain the reasons behind the department's level of funding. Fire officers should read this section and relate it to their own community. Once this is done, try to identify what the tax burden is for property owners and income or wage earners in the community. How does it compare with surrounding areas? Can the residents afford to spend more on public services? Will they support increased funding for the fire department?

This information is especially critical in small communities relying on fire levies for operating and capital funds. Decisions about what type of funding source to solicit, when to ask for funding, and how to address the concerns of the residents need to be made within the framework of the community's current tax burden. In addition, the department must to be sure to follow appropriate procedures to solicit funding legally.

Other than voluntary fundraising activities, such as chicken barbecues and bingo, fire departments have to rely on two sources of revenue: grants and tax-based revenue. In the fire service, the number of grants available from federal and state sources is extremely limited. Most of the few programs that do exist support training activities and are restricted to that use. Thus, the primary source of funds comes from the public in the form of taxes.

Local governments have the legal right to levy a charge on residents or visitors to a community in order to acquire sufficient funds to pay for services. As a department of local government, the fire department receives an allocation of funds through the budget process. A fire district is a type of independent special district. These special districts are legal local governments that also have the power to tax and to enter into legal contract. The types of taxes available to local governments are: property taxes, sales taxes, excise taxes, income taxes, severance taxes, fees, and licenses.

Property tax is the main source of revenue for local governments. Property tax is paid on buildings and real estate (land) according to their value in order to pay for local services received. There are four steps to setting and collecting property tax:

1. *The property is assessed or assigned a market value.* This is usually done by an assessor or collector at the county level. The assessment must be equitable, and centralizing the function of calculating property values helps achieve this. In addition, because a number of local governments may be taxing the same property, having only one central assessor makes the process easier. (For example, the county-based assessment may be used by the county, the city, the school district, and the water district.) This property assessment is crucial because it indicates the tax rate necessary in order to raise a given amount of revenue.
2. *The tax rate or millage is set.* A mill is one one-thousandth of a dollar ($00.001); thus, a 5 percent tax rate is a mill of 50 (50 × 0.001 = 0.05). In an incorporated city, the millage is usually set by council. In special districts, the millage is determined by a vote of the registered voters in the area. The residents of the district must approve all taxes. When the tax will be earmarked for a specific use, such as purchasing fire apparatus or building a station, many cities and other local governments require approval of the voters before the new tax can be levied. This process may require a lot of public relations work on the part of fire officers and fire fighters.
3. *The amount of tax owed per piece of property is computed.* The assessed value of the property multiplied by the tax rate equals

the tax amount. Thus, a $100,000 home in a locality with 70 mill would pay $7,000 in taxes per year.

Assessed value of property = $100,000.00
Tax rate is 70 × 0.001 = 0.07 or 7%
$100,000.00 × 0.07 = $7,000.00
What would it be on a $40,000.00 home?
$40,000.00 × 0.07 = $2,800.00

Knowing what the computed tax would be on a given piece of property is very important. A tax of almost $3,000 a year is more than many owners of $40,000 homes can afford to add to their payments; it would add an extra $233.34 each month. Thus, a levy for this amount of tax would probably be defeated. It is also necessary to look at the other taxes being placed on a piece of property to determine whether the residents can afford an additional fire levy. For example, one community has a city property tax of 20 mills, a hospital district tax of 5 mills, a water district tax of 5 mills, a fire department operating levy of 15 mills, a school construction levy of 25 mills, a school operating levy of 75 mills, and a special school levy of 10 mills. That is a total millage of 155, or 15.5 percent! A home owner would pay $6,200 per year in property taxes on a $40,000 home. Needless to say, when the fire department requested another continuing levy, the community was not very supportive.

4. *The tax is collected.* The county typically collects property taxes for all of the local jurisdictions within its boundaries. The county is usually reimbursed for this service; however, the cost to each local jurisdiction is much less than it would be for each to collect the tax separately. Most homeowners with mortgages pay the taxes as part of their monthly payment.

A **sales tax,** also known as the general sales tax, is a tax placed on items purchased within the limits of the local jurisdiction. It is termed "general" because this tax applies to virtually all items sold. In many areas food and medicine are exempt from the sales tax. In some states clothes are also exempt. State governments implement a sales tax throughout the state. Local governments who also implement a sales tax are said to "piggyback" the tax on top of what the state has set. For example, a resident might pay a 4 percent state sales tax, a 1 percent county sales tax, and a 2 percent local sales tax for a total of 7 percent of the value of any item purchased. The state government usually collects the sales taxes from businesses and then remits the appropriate share to the local governments.

An **excise tax** is also known as a selective sales tax because it is levied on specific items. This tax is also known as a "luxury tax" when placed on items such as furs and jewelry and as a "sin tax" when levied on alcohol and cigarettes. An excise tax on gasoline is often dedicated to road repair or mass transit, following the argument that persons using the gasoline should pay more for repairing the roads or assisting those who use mass transit. Excise tax is collected along with the general sales tax and then remitted to the local government.

The **income tax** is the primary source of federal government revenue. In addition, a majority of states collect individual state income tax and corporate income tax. In those states, many local governments have also enacted a local income tax. Some communities apply what is technically called a payroll tax. This tax is paid to the community in which a person works instead of where he or she lives (thereby taxing those who use the city's services on a daily basis). Thus, a person working in one city and living in another may pay a payroll tax to one and an income tax to another. Usually, communities such as this allow some offset in taxes owed. For example, a person earning $30,000 in a city with a 1 percent payroll tax will pay $300 in taxes to that city. If that person lives in another city with a 1.5 percent income tax, he or she will also owe $450 to that city. If there is an offset agreement, the city of residence could agree to collect only 0.5 percent, or $150. What is more likely, however, is that a 50 percent credit will be allowed, i.e., the worker will not have to pay 50 percent of what he or she paid to the place of employment (50% × 1% = 0.5%). Thus, the worker would owe the government of his or her place of residence only 1 percent of his or her income in taxes (1.5% – 0.5% = 1%). [Note: Most local jurisdictions have what is termed an income/payroll tax to cover both contingencies.]

Severance tax is tax placed on the removal of natural resources, such as coal, oil, or lumber, from a community. Areas with high levels of revenue generated from severance taxes are usually well endowed monetarily. However, once the natural resource is gone, the severance tax revenue, as well as many local jobs, is lost also. This can create to severe budgetary constraints.

Fees and licenses are other ways in which local governments can raise revenues. Fire departments have used fees such as EMS transport fees, inspection fees, and hazardous materials reporting fees. Fees are charged to those who use a specific local service that is not used equally by most residents. The rationale for this is that those who use the service should pay more for it.

License fees are charges by government for the privilege of conducting a given business, such as selling liquor or working in a beauty salon. Licenses are also sometimes called certification, such as EMT certification or paramedic certification. License fees are charged in order to compensate the government for the cost of regulating the quality of a service or product provided to the public.

These are the primary sources of revenue available to local governments. Fire service administrators need to be familiar with the revenue sources in place in their community. They also need to be aware of the tax load the residents are carrying. If a fire department wants to raise revenue from the public through taxation, its officers must have a good understanding of the public priorities and ability to pay.

Borrowing and Debt

Fire departments are often faced with the challenge of raising the money for making capital purchases. Many communities cannot afford to raise the necessary amount of capital all at once. The local government then seeks to do what many consumers do: go into debt. Just as people borrow to build a home or buy a car, local governments borrow to build a fire station or buy a fire apparatus. There are, however, severe limitations set on a local community's ability to go into debt. These limitations are set in state constitutions and law in order to prohibit abuse.

Local governments borrow money by issuing bonds. Essentially, the government is borrowing money from investors who buy the bonds. The investors are then repaid with interest as the bonds mature. Bonds can be repaid either in total on maturity or through partial payments made on an annual basis. The latter are called serial bonds and are generally considered safer, as they reduce the impact of the payment and remove the temptation to use the funds being collected for a different purpose.

The interest rate is determined in part by the local government's bond rating, which is similar to an individual's credit rating. Independent bond raters, such as Moody's and Standard and Poor's, examine the fiscal standing of the local government in terms of indebtedness, revenue base, demographic and economic trends, and fiscal responsibility. A high bond rating (A+++ or AAA) means there is little risk of losing money on a bond issued from that government. Lower bond ratings have a higher risk of the local government defaulting and not paying the debt. Consequently, bonds issued by governments with high ratings are less risky and have lower interest rates, while bonds from high-risk governments with low bond ratings have higher interest rates. (Most municipal bonds are also tax-free for the investors.)

In order to issue a bond to go into debt, most local governments must receive permission from their residents. Thus, they hold an election and voters have to approve the issuance of the bond. This is especially true when the repayment of the bond will require levying a new tax, although this is sometimes presented on a different item on the ballot.

There are three type of bonds, identified according to the type of guarantee each has. A **revenue bond or direct obligation bond** is guaranteed to be paid back from revenue generated by the project. For example, a turnpike is often built with a revenue bond because the toll money will pay off the debt incurred by building it. The purchase of EMS squads could be guaranteed by transport fees in departments that charge fees to either all service users or nonresidents of the community. This can be risky because if the projected income is not generated by the project, there is no money with which to pay off the bond. Thus, if the fire department only collects fees from transporting nonresidents, its ability to pay back the bond debt depends on the number of nonresident victims and the number who pay for service as billed.

A **general obligation bond** is backed by the "full faith and credit of the issuing government." It is backed by the existing revenue base of the local jurisdiction and can be paid back from existing taxes or from taxes that can be levied (and will be) if the current rates are not adequate. The general obligation bond is the most secure because the local jurisdiction *will* raise the revenue to pay it back. This type of bond usually requires voter approval. Thus, a city could issue a bond to build a fire station that would be paid for by a current property tax dedicated to capital improvements.

The **indirect obligation** bond is backed first by the revenue generated by the project and second by the full faith and credit of the local government. In this case, the revenue from the project is allocated to paying off the debt. However, if the revenue is not enough, the government will then repay the remainder of the debt. This is also a very secure bond and has the added attraction of not necessarily requiring a new tax on the public. Thus, if EMS fees were not adequate to cover the cost of the new squad, the fire district would have to take money from its operating or capital funds or the city would have to cover the debt payments from other sources.

When planning to request a fire levy, the leaders of the fire service need to be well informed about the financial status of the city and the capability of the residents to pay an additional tax. Economic and demographic trends are important considerations as well. An aging community that has lost its primary industries does not have good future revenue prospects. A fire department asking for a new levy just after the local school district has passed one is going to have difficulty convincing the voters to pay even more for government service. All of these factors must be considered when planning capital improvements and purchases.

COSTING OUT NEW PROGRAMS

When starting a new program, administrators are often confused as to how to estimate the cost to operate it. It is important for the initial estimates to be accurate in order to demonstrate knowledge of operations and ensure adequate, but not excessive, funding.

One way to begin is to contact communities that already have the same program in place. In doing so, however, it is important to contact communities similar to one's own in terms of population, demographics, service area, department size and operations, and department budget. A combination department serving a 23-square-mile community with 55,000 residents should not survey only full-paid departments in cities with more than 200,000 residents, even though they are providing similar services. The service requirements and service delivery structure are different, and costs will vary correspondingly.

As information is gathered, it is important to put it into perspective. For example, in trying to determine the salary of a full-time fire inspector, a survey of departments in communities between 25,000 and 55,000 population with combination departments reveals a range of salaries from $24,000 to $38,000 annually. The same survey reveals that part-paid fire fighter/paramedics earn from $5.00 to $10.97 per hour. Further examination of the information, however, indicates that some inspectors are paid 90 percent of the local wage rate for full-time fire fighter/paramedics. The difference is that the inspectors in these communities are not required to have fire fighter or paramedic certification. Thus, one alternative for this community is to hire an inspector at 90 percent of current department fire fighter/paramedic salary.

In a similar fashion, a good administrator can determine in advance what types of expenditures may be encountered. This enables better planning and preparation for implementing programs. Contacting only one or two departments may not provide the information necessary to maximize the resource use in one's department. The fire administrator must make an objective survey to determine the optimal method of providing a service at the lowest cost.

While salaries are generally regionally based, ideas about how to structure new programs are not. When looking for alternative methods of delivering a service, the fire service administrator should not be restricted to local or even statewide considerations. Innovations occur all over the country. The Learning Resource Center (LRC) at the National Fire Academy in Emmitsburg, Maryland, is one helpful resource for exploring alternative budgeting and funding methods.

Expenditure Restrictions

Depending on the source of the revenue funding the department, there may be restrictions attached to how funds are used. Reference has already been made to the fact that state and federal grants often limit the use of funds to specific items. The same may be true of locally generated funds. Some fire department levies are dedicated to a specific use. The department and the elected officials are prohibited from transferring these funds or using them for another purpose, even though such a transfer might be in the best interest of the community.

Fire department administrators must become familiar with available funding sources and their use restrictions. When increased funding is required for an activity not covered by existing, dedicated revenue sources, it is necessary to appeal for allocations from general funds or to identify alternative sources of revenue. A good fire department administrator is familiar with the overall operations and financial situation of the community and keeps current not only with service mandates, but also with external funding programs available.

Budgeting is not necessarily an easy task. However, good budgeting practices can be rewarding for the administrator, the department, and the community.

EXAMINATION OF INDIVIDUAL DEPARTMENT BUDGETS

To better understand and relate to the information presented in this chapter, we suggest you examine your department budgets and answer the following questions:

Does your department have both an operating and capital budget?

What is the total amounts of each?
What is the largest category of expenditures for each?

For the Capital Budget:

Is it a 1-year or multiyear budget?
What items have priority in terms of purchasing allocation?
Do these purchases relate to specific programs or general operations?
What is the source of revenue for the capital budget? (Is it a special levy? Is the revenue restricted to capital purchase?)

For the Operating Budget:

What budget format do you use?
What is the budget cycle for your department? (You may need to talk with the chief to find this information.)
Have there been any significant changes in expenditures over the past 2 years?
Do these changes correspond with changing priorities in the department?
Have there been any resource shifts among programs in the department?
What are the sources of revenue for the department's operating budget? (If it comes from the "general fund," where does the general fund get its revenue?)
Has there been a change in the nature of the revenue source or tax base in the community over the past 5 years (new developments, plant closings, annexation, etc.)?
How might these changes impact the revenue available for the department in the next 2 to 3 years? Will there be a corresponding shift in service demands?
How do the department planning activities interface with the budgeting process?
What improvements in budgeting would you suggest for your department?

SUMMARY

Budgeting is one of the most important aspects of administration. Without the proper allocation and use of resources, a fire department can not function to its full capacity. Wasting public resources is a violation of the public trust—a misuse of money, which is becoming increasingly scarce.

Fire administrators must learn how to budget well. They cannot simply accept funds and say they need 10 percent more than last year. Examining the allocation and expenditure of resources forces the examination of the operation of the department as a whole. Today and in the future, the successful administrator must know exactly how resources are used and why.

Budgeting is a complex activity of tremendous importance. Useful and responsible budgeting matches resource allocation with the service needs of the community and the plans of the fire department. It is the activity that facilitates the fulfillment of plans.

ENDNOTES

1. *De facto* means "in fact" and is used when something is not done by law or official proclamation, but simply takes place. For example, if a department's short-range plan contains a Sparky presentation of fire safety materials at the mall during Fire Prevention Week, but no money is allocated to pay for the fire fighters to go to the mall or purchase the materials, the "official" policy or plan is to have the program; *in fact,* however, the program has been canceled.
2. The fiscal year is a 1-year or 12-month period that identifies the point in time at which the budget is approved. The fiscal year may or may not coincide with the calendar year. Many begin on July 1 and end on July 30. The federal fiscal year begins October 1 and runs through the following September 30.
3. Robert B. Denhardt, *Public Administration: An Action Orientation* (Pacific Grove, CA: Brooks/Cole Publishing Co. 1991), p. 170.
4. The priority rating matrix is adapted from the presentation in Sara Hendricker, *Municipal Budgeting in Ohio* (Columbus, OH: The Ohio University and the Ohio Municipal League, 1986), pp. 1–4.
5. NFPA 1500 might require that the fire department obtain additional equipment, but it does not necessarily mean it must replace existing equipment.

REFERENCES

Bozeman, Barry, and Stuart Bretschneider. 1986. "Public Management Information Systems: Theory and Prescription." *Public Administration Review* 46. Special Issue (November). pp. 475–487.

Bubankis, Michel. 1976. Budgets: *An Analytical and Procedural Handbook for Government and Non-Profit Organizations.* Westport, CT: Greenwood Press.
Denhart, Robert B. 1991. *Public Administration: An Action Orientation.* Pacific Grove, CA: Brooks/Cole Publishing Co.
Graham, Cole Blease, Jr. and Steven W. Hays. 1986. *Managing the Public Organization.* Washington, DC: Congressional Quarterly Press.
Hendricker, Sara. 1986. *Municipal Budgeting in Ohio.* Columbus, OH: The Ohio University and the Ohio Municipal League.
Lee, R. D., Jr. and R. W. Johnson. 1983. *Public Budgeting Systems,* 3rd ed. Baltimore, MD: University Park.
Mikesell, John L. 1982. Fiscal Administration: *Analysis and Applications for the Public Sector.* Homewood, IL: Dorsey Press.
NFPA 1021, *Standard for Fire Officer Professional Qualifications,* 1992 ed. Quincy, MA: National Fire Protection Association.
NFPA 1500, *Standard on Fire Department Occupational Safety and Health Program,* 1992 ed. Quincy, MA: National Fire Protection Association.
Pagano, Michael A., and Richard J. T. Moore. 1985. *Cities and Fiscal Choice: A New Model of Urban Public Investment.* Durham, NC: Duke University Press.
Rabin, Jack, and T. D. Lynch. 1983. *Handbook on Public Budgeting and Financial Management.* New York: Dekker Publishing.
Steisss, Alan Walter. 1989. *Financial Management in Public Organizations.* Pacific Grove, CA: Brooks/Cole Publishing Co.
Wanat, John. 1978. *Introduction to Budgeting.* Monterey, CA: Brooks/Cole Publishing Co.
Williams, J. D. 1980. *Public Administration: The People's Business.* Boston, MA: Little, Brown and Co.

CHAPTER 12

Ethics

This chapter presents a general discussion of ethics and then focuses primarily on the challenges facing the fire service. All fire officers need to understand the concept of ethical dilemmas and the importance of dealing with them appropriately. There are few absolutes in matters of ethics. It is a gray area in which a right or wrong answer is not always apparent. This may present a challenge to fire officers, but the importance of maintaining high ethical standards in the fire service must not be overlooked.

The subject of ethics has become a recurrent theme in discussions of leadership in all aspects of public service. Many experienced fire officers recognize the need for correct ethical behavior, but no consistent approach to resolving ethical dilemmas has emerged. Recent scandals associated with prominent public officials have caused citizens to be more skeptical of public administrators and public service agencies, including the fire service. In recent years, some citizens groups have demonstrated a much more critical view of traditional municipal services. The tradition of broad public acceptance of the fire service as an agency existing only to provide service can erode quickly if a high standard of ethical behavior is not maintained by all department personnel.

The National Commission on the Public Service concluded in its 1989 report, "If government is to be responsive to the people's will and capable of meeting the challenges of the twenty-first century, it must have a public service of talent, of commitment, of dedication to the highest ethical standards."[1] Professor Carol W. Lewis suggests that "ethics and genuine professional success go together in the ethical enterprise called public service."[2] There is, however, no clear consensus about what constitutes ethical behavior. Furthermore, the question of what ethical behavior means for public administration is also vague.

The fire service, as a public service organization, must address the issue of ethical conduct. As a public administrator, the fire officer should recognize the impact of ethical behavior and set the tone to ensure its presence in all aspects of the agency's operation.

The framework to mold ethical awareness is in place. Numerous fire department mission statements contain the phrase, "to serve the public." Recognizing ethics and accepting ethical behavior as the expected standard of conduct brings about the highest level of service possible. This is because ethical behavior involves fulfilling the trust placed in one by another. Thus, the public's trust in the fire service to provide the best protection possible presents an ethical standard requiring maximum performance by all members of the fire department in all aspects of the job, both in daily operations and emergency response. The implied ethical acceptance of the public trust should be articulated in the mission statement or primary goals of the department by outlining principles for ethical behavior and a statement about upholding the public trust.

WHAT CONSTITUTES ETHICS?

What is ethical behavior and how is it defined? **Ethics** are standards or principles of conduct that govern the behavior of an individual or group of individuals.[3] Moral standards are usually translated into some form of ethical conduct generally concerned with what is right and what is wrong. For example, an Ohio Supreme Court Justice stands accused of voting on cases involving a bank in which he owned 900 shares of stock worth almost $20,000. In this instance, the justice failed to disqualify himself from any of the seven cases and never voted against the bank. The justice also, on three occasions, voted favorably for a corporation while owning stock in that company.[4]

To most citizens, this action violates the trust placed in persons holding public office. It further demonstrates that responsibility and accountability cannot be avoided, even at the risk of personal loss. The authority entrusted to such a position should translate into ethical behavior and permeate official matters.

Situational Ethics

Moral standards are uniform and are applied in the same manner in each situation. Something that is wrong or unethical is wrong or unethical all of the time, in every situation. **Situational ethics** is a term applied to standards that change according to the situation at hand, much like situational leadership encourages leaders to do. (See Chapter 2, Leadership.) However, changing standards according to the situation is unacceptable in ethical matters because the situation is used to excuse unethical behavior. For example, someone who takes only "a little" (less than $200) from wealthy individuals and then uses the money to buy food for a homeless family might justify stealing on the basis of neediness. The premise is that under certain circumstances, or situations, stealing is an ethical act. Stealing is acceptable if the victims can afford it and the money is used for a "good" cause.

Opponents of situational ethics do not accept excuses for unethical behavior. Stealing is wrong whether it involves taking things from rich people who are insured or whether it involves taking money from someone who needs it to buy food. Stealing is wrong, no matter what the situation.

A variation of situational ethics is the concept of **utilitarian ethics,** in which ethics are employed when they are useful or serve a purpose. Certain actions or values are considered ethical when convenient or in certain contexts. However, when ethical choices are inconvenient, the ethical criteria for a choice change. **Nonutilitarian ethics** relates to a more absolute acceptance of right and wrong. An act that is wrong is wrong regardless of the degree involved. Thus, stealing $2 from a store is as wrong as stealing $2,000 from a retired person living on social security.

Consider the following example of typical ethical dilemmas. These are the types of ethical challenges fire officers are likely to face during their career. As you think about the situation, remember the concepts of public trust and situational ethics.

Case Study #1

A week before Thanksgiving, the fire department rescues a 17-year-old from a burning vehicle. After the incident has been mitigated, the fire fighters on the scene are informed that the young man is the son of a prominent local automobile dealer. The dealer is so grateful for the rescue of his son that he donates one turkey to each member of the department the next day.

1. Is it ethical to accept the turkey?
2. Have grateful citizens dropped off boxes of candy, cookies, and cakes at your fire station for the enjoyment of the department members?
3. As a department member, would you accept the turkey?
4. As chief, how would you react to the offer?
5. As an elected official (mayor, council person, trustee), how would you react to the situation?

Case Study #2

A week after the incident involving the rescue of the young man, one of the fire fighters who extricated him from the burning vehicle and accompanied him to the hospital decides to purchase a new personal vehicle. The fire fighter recalls meeting the young man's father at the hospital and decides to visit his dealership. The automobile dealer recognizes him and offers to sell the fire fighter a new car at cost, with all accessories paid for by the dealership. This would result in a savings of over $2,000 to the fire fighter.

1. Should the fire fighter accept the offer?
2. Does the dollar value of the gift determine whether it is ethical to accept?
3. Is it acceptable if everyone in the department benefits equally, but unacceptable if only one individual benefits?

4. Did the fire fighter instigate the car deal?
5. As a department member, how would you feel about the deal?
6. As chief, how would you react to the offer if you heard about it (which you eventually would)?
7. As an elected official (mayor, council person, trustee), how would you react to the situation?

These questions present ethical dilemmas. Most would agree that the car deal is unethical, or at least unfair. However, most would agree that accepting a turkey at Thanksgiving is appropriate, especially if everyone in the department gets one. This is seen as a definite expression of gratitude, as opposed to a payoff, and it makes everyone "feel good." But is it ethical? No.

Although situational ethics would argue that the gift of turkeys was ethical whereas the car deal was not because the situations were different, other ethics experts argue that situational ethics is a way to excuse unethical behavior. Thus, both situations would be unethical. Who benefited, how the gift was made, and the value of the gift do not make one situation ethical and the other unethical.

Sometimes the ethical dilemma involves selecting the lesser of two evils. The situation is such that regardless of which course of action one takes, one is violating personal values.

Consider for a moment this dilemma. You come upon a man on the street who is suffering a potentially fatal medical condition. There is a drug store just across the street, but it is closed. Because of your previous training, you know which drug to administer to save this person's life. The ethical dilemma is: Do you break into the drug store, administer the drugs, and save the person's life or do you make the victim as comfortable as possible while he dies, because it is unethical and illegal to break into the drug store and take the necessary medication?

Often, these are the only two options that come to mind. There is, however, a third course of action. You could break into the drug store, administer the drugs, and then call the store owner and offer to pay for damages and for the medicine. This course of action attempts to reconcile or ameliorate the violation of ethics.

Consideration of Cultural Differences

Ethics are a result of the social forces and human experiences accumulated over hundreds of years.[5] As such, ethics are closely tied to the culture of an area, and as cultures differ from place to place, ethics may also differ. This is distinct from situational ethics. Once it was culturally ethical to restrict women to the position of dispatcher in emergency agencies. With changes in cultural values, however, denial of equal opportunity is today not only illegal, but also unethical. Ethics reflect the prevailing culture of a community and an organization. What may be considered unethical in one geographic area of the country may not have the same connotation in another because of cultural differences.

In communities with changing demographics, many new cultures can be found. In these communities behavior that seems unethical or immoral to Americans may be the accepted practice in the culture of origin. For example, in several African countries the practice of giving a gift to a judge or someone who will make a decision to resolve a dispute is part of the culture. The idea behind the gift is that the individual is thanking the judge for solving a problem. This type of activity would be considered a bribe by U.S. law. In addition, it is considered unethical according to our cultural value system.

Thus, when dealing with individuals of different cultures, fire officers should be aware of various customs. When confronted with potentially unethical situations, it is best to explain why the situation is considered unethical by American standards or those of the fire department. Being exposed to cultural differences is not an excuse to violate one's existing ethics, but it does require an awareness of various ethical standards so that the impact of conflicting cultures can be eased.

Manager Ratings of "Degrees" of Ethics

In 1975, William A. Ruchan and John W. Newstrom published the results of a study conducted to determine what actions or activities were considered unethical by managers and supervisors.[6]

Participants were asked to rank seventeen actions (items) in terms of how unethical they considered each to be. It is not surprising to note that embezzlement and stock fraud were universally ranked as being unethical. Researchers noted that taking advantage of an innocent peer was seen as being most unethical. However, there was a considerable difference of opinion over how unethical other specific activities were.

The items were ranked in descending order and placed into three categories, which included items judged as most unethical, not as unethical as the first group, or least unethical.

Items Judged as Most Unethical

1. Passing blame for errors to an innocent coworker
2. Divulging confidential information
3. Falsifying time/quality/quantity reports
4. Claiming credit for someone else's work
5. Padding an expense account more than 10 percent
6. Pilfering company materials and supplies

Items Judged Not as Unethical as the First Group

7. Accepting gifts/favors in exchange for preferential treatment
8. Giving gifts/favors in exchange for preferential treatment
9. Padding an expense account up to 10 percent
10. Authorizing a subordinate to violate company rules
11. Calling in sick to take the day off

Items Judged as Least Unethical

12. Concealing one's errors
13. Taking longer than necessary to do a job
14. Using company services for personal use
15. Doing personal business on company time
16. Taking extra personal time
17. Not reporting others' violations of company policies and rules.[7]

The survey conclusions represent only averages, not consensus, since many of the 121 managers and supervisors differed on the degree of severity of each unethical activity.

Several of the unethical actions listed in the three categories may also be violations of the law in most states. Others violate either professional codes of ethics or agency rules and regulations. Although each of the unethical actions has been categorized, the actual interpretation of the degree of seriousness of each remains with the authority having jurisdiction. Cultural differences and local attitudes toward ethical conduct may result in some shifting among the degrees of seriousness, to adjust priorities and to align them more closely with local acceptable standards.

This study illustrates the fact that ethics is not always a clear-cut issue of right and wrong. However, note that the headings of the three groupings were "most unethical," "not as unethical," and "least unethical." No item was grouped into a category titled "ethical." Thus, when dealing with degrees of unethical behavior, it is important for managers to remember that it is all *unethical* behavior. Even "slightly unethical" actions are unethical and, although the severity of reaction or discipline should be adjusted to the severity of the misbehavior, the administrator should never simply ignore unethical actions. That is an implicit endorsement of lack of ethics and leads to the question, Where do you draw the line?

Ethics is a central part of each of us. However, since it is something very personal and individual, it has not been dealt with on a formal basis often enough.

ETHICAL ACCOUNTABILITY AND RESPONSIBILITY

Fire officers must be prepared to address ethical conduct in the fire service. For example, it is no longer ethical to conclude that the sole mission or purpose of a fire service organization is to save lives and property. The conditions under which these tenets were founded have changed dramatically. As such, the fire administrator must reassess the mission of the agency and make ethical decisions concerning the *accountability* and *responsibility* of the organization.

The total number of fires and associated fire fatalities diminished in many communities during the 1980s. Still, many fire officers are unwilling to acknowledge these facts. As an example, a fire administrator who continues to demand more personnel, equipment, and other municipal resources without making an effort to (1) improve the current level of service, (2) introduce new services, and (3) consolidate resources may soon be confronted with the ethical issues of accountability and responsibility. In this instance, failure to recognize and react to changing regional conditions will cultivate an ethical problem that could negatively affect the fire service agency and the local government entity it serves.

The issue of ethical behavior should permeate all levels of the fire department. No bureau or division is immune from operating under ethical standards. It begins with recognizing what constitutes correct ethical behavior at the local community level. "Public managers are responsible for the actions undertaken under the umbrella of their authority and in their name."[8]

Accountability requires that when unethical behavior occurs, the individuals involved are reprimanded and required to compensate for their behavior. Excusing unethical behavior is almost as bad as identifying unethical behavior and then ignoring it. Ignoring acknowledged unethical behavior simply teaches that ethics are not important and do not matter in the fire department.

Fire service officers must recognize the symptoms of unethical conduct and learn the steps required to successfully identify both the root and symptomatic elements of such behavior. Read the following case studies and decide how you would respond to the situations. What changes or actions would you recommend?

Case Study #3

Bud Baxter is the senior captain of a volunteer fire department located in a moderate-sized community. He has served on the department for 11 years and is known for his prowess at emergency scenes. Three years ago, the chief assigned Bud the additional responsibility of scheduling and conducting all department EMS and fire training at the suggestion of the other officers. Bud is respected by the members of the department and has considerable experience.

One Sunday evening in late fall, the department is dispatched to a reported vehicle fire on a rural lane in a remote location of the community. Over the past 2 years, the department has been called to numerous vehicle fires in the same area. While responding to the incident, Bud remarks to the apparatus driver that "this is probably another stolen vehicle that has been torched." Arriving on the scene, they discover a late-model pick-up truck fully involved in fire. Bud has correctly assessed the situation and directs the fire fighters in extinguishing the blaze.

After the fire is extinguished and the overhaul is complete, one of the fire fighters notices Bud examining the contents of the pick-up truck bed. Within a few minutes, Bud returns to the apparatus with a large tow chain and several hand tools that he has retrieved from the burned vehicle. When questioned by other members as to why he has placed this equipment in a rear compartment of the fire department apparatus, Bud replies that he needs the chain and tools for his tree-trimming business. Besides, "the insurance company will cover the loss for the owner." Bud assures the other fire fighters that this is no big deal and that he has done it on several other occasions involving vehicle fires and motor vehicle accidents.

1. Did the captain demonstrate appropriate ethical conduct?
2. What is the impact of his actions on other members of the department?
3. What are the ethical responsibilities of the other fire fighters on the scene who witnessed Bud's actions?

It is obvious that Bud Baxter's actions were unethical. In fact, the removal of contents from a vehicle or structure is not only unethical, it is also a violation of the law. Under these conditions, fire officers must be prepared to confront the personnel involved *immediately.* Failure to do so clearly demonstrates a lack of fire officer leadership. It also sends a message to other department personnel that actions such as this are condoned by department hierarchy as well as local elected officials.

It is not true that no one is hurt in the above situation. Although the owner of the vehicle will be reimbursed, other persons insured by the same company are actually paying. Greater damage is done to the members of the fire department and to the fire department itself. Baxter's action told the fire fighters that it is acceptable to steal from an insurance company, but not from an individual. In addition, he demonstrated that unethical behavior in one instance is not only accepted, but almost encouraged. If it is alright in this situation, why not in another? This is an example of an application of situational ethics. It also illustrates the difficulty of defining boundaries when situational ethics are followed. Baxter's accountability as a fire officer was compromised because he did not follow accepted standards of ethical behavior.

The ethical responsibilities of other fire fighters are viewed differently among members of the fire service. The exact nature of ethical responsibility of fire fighters on the scene is uncertain; however, research demonstrates that most officers would agree that something should be done, ranging from confronting Bud to reporting him to his superior.[9]

Case Study #4

Noland Township Fire Department maintains three stations in a rapidly expanding suburban community. It has achieved and maintained a positive image with both the news media and citizens by providing efficient emergency response services and implementing effective fire inspection and public education programs. Once operated by volunteers from the area, the department's conversion to a combination department (both full-time and volunteer or on-call personnel) occurred 11 years ago due to a dramatic increase in the population and corresponding emergency calls. The main station is manned at a minimum level around the clock by ten full-time fire/EMS-trained persons. The two satellite stations are manned by paid on-call personnel who respond as needed in an emergency situation or for required training.

A number of the paid on-call personnel frequently visit the main station in anticipation of responding on emergency calls with the full-time staff. They are not compensated for this time unless a call for other paid on-call personnel is dispatched. Quite often during slow periods, some department personnel (both full-time and on-call) move apparatus out of the station and wash their personal vehicles in the vacant bay. The mayor has questioned the fire chief about this routine several times during the past year. The chief has defended the practice as a fringe benefit offered to all fire department personnel. He stated that the practice is a carryover from the days of the all-volunteer department, when these "perks" were offered in exchange for the member's service.

On a cool Saturday afternoon, Jamey, a paid on-call member, moves his personal vehicle into the department's maintenance bay and repairs an apparent electrical problem in the engine compartment. During the process, he uses department resources, including: (1) electrical analyzing equipment, (2) numerous hand tools, (3) a number of new screws and bolts stored in the supply cabinet, (4) electrical tape and wire connectors from the supply cabinet, and (5) several feet of new electrical wire stocked by the department for use on emergency response vehicles.

When the chief asks Jamey about performing the work on his personal vehicle, using department equipment and resources, Jamey replies that he has observed other department personnel doing it, and assumed he was entitled to do the same. As an example, he notes that recently he assisted a full-time fire fighter who was repairing a snowplow bracket on his personal vehicle using the department's welding equipment and associated welding supplies.

The chief reminds Jamey that use of department equipment and resources to repair personal vehicles is not a standard department practice. Jamey replies that he feels the "couple of bucks" worth of supplies used were in return for his countless hours of hanging around the main station and assisting the full-time fire fighters with their maintenance chores.

In this situation, the fire officer must consider the ethical implications of the members' actions. Consider the following questions.

1. Is washing of personal vehicles by department members, using department resources, a violation of ethical standards consistent with your local environmental culture?
2. Does the use of department equipment and resources for personal benefit constitute inappropriate ethical behavior?
3. Is the fire department obligated to provide financial rewards or other benefits to persons who choose to "hang around" the station even though they are neither scheduled to work nor summoned to an emergency situation?
4. Did the fire chief's defense of washing personal vehicles contribute to the use of department equipment for personal vehicle repairs?

5. Should Jamey be formally reprimanded for his actions? Should he be required to pay for the supplies used?
6. Should the fire chief address the issue of the full-time fire fighter repairing his snowplow? Why or why not? If so, how should he approach the issue?

This case study highlights the need for consistent ethical correctness. It should be examined to determine how much unethical behavior has been displayed and its prevalence throughout the department. As you reconsider the scenario utilizing these guidelines, ask yourself the following questions:

1. In your community, what is the public attitude toward fire department personnel using department space and resources to wash personal vehicles at no cost to the member (positive or negative)?
2. Is the use of fire department equipment and resources to repair personal vehicles in line with the ethical standards established by your department rules and regulations or legislation adopted by local elected officials?
3. Is this fringe benefit available to all other municipal employees, including law enforcement, road, water, and sewer (regardless of employment status: full-time, paid on-call, or volunteer)?
4. Is this free benefit available to all citizens of the community?
5. Have any professional codes of ethics, rules, or laws been violated by such action?

The case study poses a question of the fire department's ethical accountability to the public. Using fire department space, water, and cleaning agents to wash personal vehicles lies within the category of actions judged as "least unethical," as discussed previously. However, the use of other department equipment and resources of greater value becomes a more serious ethical issue. In many states, legislation prohibits the use of resources from tax revenue allocations for personal benefit.

In your own personal situation, the use of department resources for personal benefit may be within the accepted ethical norm for the department and the culture of the community. An analysis of the department's code of ethics, or rules and regulations, along with local and state legislation should help guide the discussion concerning ethical conduct in your community.

ESTABLISHING STANDARDS FOR EVALUATING ETHICAL CONDUCT

How does a fire officer define appropriate ethical behavior for the department? Because of the cultural diversity found in many communities, attempts to clarify ethical standards must begin with an understanding of the local culture. What is and what is not acceptable ethical behavior to department personnel and the citizens of the community? Is there a local or state law that defines the guidelines for ethics? Does the local jurisdiction support a "code of ethics" for all municipal personnel regardless of paid or volunteer status?

The following questions can be used to construct the framework for a comprehensive study to define the boundaries of ethical behavior. Assess your personal thoughts about ethical standards and conduct as you answer.

1. When situations occur that test your personal ethical convictions, do you react to correct them?
2. Do your actions and decisions "feel good" to you?
3. Do your actions meet the perceived level of acceptable ethical conduct for your department?
4. Do you have trust in yourself and in your judgment? Are you questioning your actions?
5. Are your actions viewed as acceptable ethical behavior by your supervisors, peers, subordinates, and citizens of the community?

The following excerpt from the article "Bye, Bye Birdman" by John A. Granito is an example of the actions fire departments must take to establish ethical standards. This case presents a number of unethical actions. The use of on-duty time and telephone numbers to run one's own business is something that is easily seen as unethical. Or is it? Many fire fighters have "B" jobs. Where should their allegiance lie? Is it okay to work on the aspects of a second job when on duty as long as it doesn't interfere with one's duties as fire fighter? How is "interfere" defined? If interference is something that involves being physically removed from the station or limits one's activities, then obviously interference is not acceptable. However, if interference is defined as anything that decreases one's mental attention and preparedness or distracts one's concentration from the job at hand, the question is not as easy to answer.

Case Study #5 [10]

I've had problems with members once in a while because of second jobs. Before I became chief, two fire fighters were caught using city-owned paint to outline commercial parking lots as part of a little business they ran on the side. Other than that, however, my biggest objection to the scheduling arrangement was that I have trouble getting people who are on 24-hour schedules to take 40-hour-week daytime positions.

Last Friday night, my wife told me she had ordered a $30 birdhouse for the backyard. She had seen an advertisement in the newspaper that said: "Fill your yard with beautiful birds. Call 444-2438 and ask for the Birdman. Handmade birdhouses. $XXX delivered."

Something about that ad—besides the price—bothered me all weekend. Yesterday I realized that the phone number was the line to our training division. When I checked it out, I discovered that the Birdman is one of our training people on a 40-hour week. When he is out of the office, the secretary takes the caller's number, then the Birdman calls back to take the order. Unbelievable!

I went right to the training chief. His excuse was that he'd never heard about it. When he questioned the secretary, she said she thought the fire fighter's nickname was Birdman. She also said that all she ever did was take phone numbers for return calls or switch calls to the member if he was in. She said she never got more than three calls in one day, and many days she didn't get any.

When the training chief called the fire fighter in to talk to him the member said that almost everyone had a second job, that he had taken the job only because the training chief had talked him into it. He said he made the birdhouses at night and that his child delivered them on weekends.

How should I handle this?

(a) Tell him to change the ad, inserting his home phone number and the words "after 6 p.m."
(b) Tell him to change the ad, and transfer him back to station duty.
(c) Discuss the problem with the local union president and ask him to straighten out the situation.
(d) Let the training chief handle the whole business so that if the fire fighter decides he is being mistreated, official grievance procedures can be followed.

While this may sound like a simple case, don't be fooled. What appears to an administrator to be clearly wrong may seem reasonable to line personnel. Many fire fighters receive personal calls while on duty, and apparently they don't interfere with the performance of their duties.

In addition, there are questions about whether the training chief knew what was going on and simply ignored it, and whether the secretary who took messages for the fire fighter had a free birdhouse or two in her own backyard.

After having a careful talk with the training chief about his administrative responsibilities, let him handle this as option D suggests. Options A and B would get you into the action too soon and would foul up any further procedures. By involving the local's president, answer C takes the problem out of the department and ignores the fact that the fire fighter has a supervisor.

A helpful rule of thumb is to remember that typically at least half your members will not side openly with the administration in a personnel matter. If you expect the worst you usually won't be disappointed. Someone once said, "Hope for the best, but plan for the worst."

Appropriate performance standards would hopefully prevent similar situations from arising. If they do occur, however, these same standards can be used to address the problems. They are also guidelines to identify what is acceptable and unacceptable.

Internal and External Controls

To ensure responsible conduct within public agencies, the maintenance of internal and external controls is necessary. **Internal controls** encompass the professional standards and values held by persons in public service. They are generally cultivated and nurtured through professional associations, such as the National Fire Protection Association (NFPA) and the American Society for Public Administration (ASPA). Special sessions and panel discussions designed to highlight ethical standards are often included in their annual meetings. Education modules are also developed for in-service training programs offered by law enforcement and other public service agencies. The purpose of the internal control mechanism is to instill in the public servant an understanding of values and ethical standards that should be employed in the absence of rules or legislation.

External controls are generally expressed in the rules and regulations of an agency or in local legislation. These laws often come into being as an external response to violations of ethical conduct. Legislation is passed in order to avoid such problems in the future. The second most common recourse is to create new organizations or restructure existing ones to establish more manageable control and monitoring systems.[11]

For example, in 1992 members of the U.S. Congress came under sharp criticism for what citizens and the news media perceived as significant violations of ethical conduct. These allegations involved uncontrolled overdraft privileges at the House Bank and certain other "perks," such as a free prescription service and reduced membership fees to the health club. Senator David Boren of Oklahoma was quoted as saying, "The banking problem . . . really is symptomatic of a broader institutional need for restructuring."[12]

To address the problem, the Speaker of the House instituted several changes, which included: (1) disbanding the operation of the House Bank, (2) securing the resignation of the sergeant-at-arms who managed the House Bank, (3) eliminating the free prescription privilege, and (4) increasing the health club annual membership fee. Thus, external controls—new rules and modification of existing organizational structure—were used to correct these ethics problems.

An example of how both internal and external controls can come into play when dealing with ethical dilemmas is illustrated in Granito's case study, "A Touchy Situation." As you read this case, answer the following questions:

1. What are the internal controls?
2. What are the external controls?
3. Who was accountable for not formally reporting the allegation?
4. Identify *all* actions that were unethical. Rank them in terms of most unethical to least unethical. Why do you classify them this way?
5. How could this situation have been avoided?

Case Study #6 [13]

As the deputy chief with administrative responsibility for emergency medical services, you are facing a problem.

Here's the story: About a week ago, a dispatcher took a 911 call from a 30-year-old woman who said she was having an adverse reaction to some medication. Her throat was swollen and her eyes were watering. She felt warm and dizzy, and her skin was somewhat red. It was 8:30 p.m. on a Friday evening, and she was alone in the house.

Because Ambulance 3 was at a local garage out of service with a broken fan belt, Medic 1 was dispatched. That left two other units in quarters. The responding unit reported that the woman's reaction was slight, but that they were transporting her to an emergency room at her own request. Medic 1 reported that it was en route to the hospital at 8:47 p.m.

At 8:48 p.m., Medic 2 was dispatched to a call for a man with severe chest pains. Two minutes later, a police cruiser reported the following multiple vehicle collision on Main Street: "Bus versus two autos; several injured; persons trapped in vehicle."

The dispatcher ordered Engines 1 and 5, Medic 4, a heavy rescue unit, and the duty chief to the scene. In addition, Medic 1 was diverted to the scene from its hospital response. A police car was to pick up Medic 3's two crew members at the garage so that they could put the reserve ambulance in service.

At the scene of the accident, Engine 1 stretched a protective line and the heavy rescue crew began to open up the car. Engine 5 was told to stand by at the corner. The crews of Medic Units 4 and 1 began to survey and stabilize the injured persons. A fire fighter from Engine 5 was told to watch the patient in Medic 1.

When the woman finally arrived at the emergency room, she told one of the regular crew members that "he had bothered her in the back of the ambulance." Though that statement never made the run report, you heard about it from the station officer a day or two later.

This morning, four days after the run, the local newspaper reported that a woman who was being transported to the hospital in a fire department ambulance "felt a hand being pushed between her legs" during the period the ambulance was stopped for a long time. Her attorney has filed charges of "criminal sexual conduct" by an as-yet-unidentified person.

The chief has ordered you to his office immediately. How do you handle this?

(a) Act as though it's news to you.
(b) Blame the company officer and the crew of Engine 5 for whatever might have happened.
(c) Tell the truth: that you had heard there was a complaint and that you will conduct an investigation.

In this case, there isn't much the supervisor can do except admit that he didn't follow up on what he had heard and then assure the chief that he will investigate.

The use of internal controls, such as codes of conduct, could have mitigated the subsequent use of external controls, in this case a criminal lawsuit. It is important to instill internal controls and to make all department members aware of external controls as well. Failure to define and monitor internal controls can lead to imposition of external sanctions on the entire department and department members other than those who are directly guilty of unethical conduct.

Legislation

The ethical values of trust, public confidence, and faith in public service employees and institutions are the fundamental elements of democracy. As such, responsible for identifying and monitoring the appropriate standards of ethical conduct lies with the public administrator.

Legislation aimed at ethical behavior can be found at local, state, and federal levels. For example, a majority of the states have adopted legislation to identify specific boundaries for ethical conduct by public service employees. The common threads found in a majority of the states' legislation include: (1) conflict-of-interest restrictions, (2) financial disclosure requirements, and (3) post-employment prohibitions. Ethics boards and commissions have been instituted in many states to administer the laws.[14]

The first federal legislation passed in 1853–54 dealt with conflict-of-interest and other abuses. Since then, legislative action has been applied to a potpourri of ethics problems at the federal level. In 1965, President Lyndon Johnson issued a Presidential Order on Ethics prohibiting government officials from holding "direct or indirect financial interests that conflict substantially with their responsibilities and duties as federal employees." President Jimmy Carter signed the Ethics in Government Act in 1978, which established detailed financial disclosure provisions and post-employment restrictions. It also established the Office of Government Ethics under the Office of Personnel Management and recommended changes in ethics rules, legislation, and policy.

Congress passed the Government Ethics Reform Act in 1989, and it was signed into law by President George Bush. The provisions of the law tightened the restrictions on representing private parties before government agencies and placed limits on the dollar amount members of Congress were entitled to accept as gifts.

Codes of Ethics

A code of ethics is a written statement of the principles that should be followed in the conduct of business. These codes are typically designed to provide general ethical conduct guidelines, address employee concerns such as acceptance of gifts and meals, support purchasing policies, and provide guidelines for employee involvement in political and nonpolitical campaigns. Organizations such as the American Society for Public Administration (ASPA), the International City Managers Association (ICMA), the National Education Association (NEA), and the National Fire Protection Association (NFPA) have taken an aggressive approach to codes of ethics. The ICMA model also contains accompanying "Guidelines for Professional Conduct and General Policy" and "Rules of Procedure for the Committee on Professional Conduct," which provide the details for interpreting and enforcing the code.

In some instances, the code of ethics statement addresses only service delivery issues or employee conduct. Other issues, such as employee privileges and gift acceptance policies, are contained within the department rules and regulations.

The National Fire Protection Association has examined the issue of ethics in the fire service and concluded that it is an important topic for the decade. A statement addressing ethical conduct was included in an informational appendix in the 1992 edition of NFPA 1021, *Standard for Fire Officer Professional Qualifications.*

> *Fire officers are expected to be ethical in their conduct. Ethics implies honesty, doing what is right, and performing to the best of one's ability. For public safety personnel, ethical responsibility extends beyond one's individual performance. In serving the citizens, public safety personnel are charged with the responsibility of ensuring the provision of the best possible safety and service.*
>
> *Ethical conduct requires honesty on the part of all public safety personnel. Choices must be made on the basis of maximum benefit to the citizens and the community. The process of making these decisions must also be open to the public. The means of providing service as well as the quality of the service provided must be above question and maximize the principles of fairness and equity as well as those of efficiency and effectiveness.*
>
> —NFPA 1121, Appendix A-1-3

The statement in NFPA 1021 is intended to address ethical standards within the context of personal conduct and service delivery to the community. Although it is provided in the document for information purposes only, it is the first such statement to appear in NFPA professional qualifications standards. It can easily be adapted for use as a generic ethical guidance document or consolidated with appropriate agency rules and regulations to form an official code of ethics document for the department.

Adopting a code of ethics improves the whole public service sector and establishes specific boundaries for ethical conduct. Whether developed within the organization or adopted from a recognized professional standard, it is an ethical road map that provides guidance and direction to members of the department.

OTHER AREAS OF ETHICAL CONDUCT

To develop a practical awareness of what constitutes proper ethical behavior, a fire officer must be cognizant of diverse situations and potential influences on ethical choices.

Loyalty

Over time, most of us tend to lose some enthusiasm for the job we have or the avocation we enjoy. As this occurs, our loyalty wanes and our interest diminishes. Part of being a fire officer is understanding that loyalty to the organization is paramount to its success. As a leader, the officer must demonstrate proper ethical conduct by viewing the organization in a positive manner. Failure to do so can undermine the organization and create morale problems at all levels. Subordinates deserve sound ethical leadership in all dimensions of fire department operations. Loyalty to the organization means: (1) accepting responsibility, (2) cooperating with others, and (3) being accountable for one's actions.

Organizational Politics

Merit and outstanding performance are the methods most often used to better oneself or gain favor in the organization. Organizational politics, however, include trying to achieve a competitive edge over one's peers by attempting to influence the chief or trying to gain power. Usually, self-interest is the motivating factor behind organizational politics. Before becoming immersed in the game of organizational politics, the officer must be certain the actions will not harm others or the organization and that they fall within the established guidelines for correct ethical behavior.

Unethical Behavior on the Part of Superiors

What should one do when confronted with unethical conduct by a superior? There are three options available in this situation:

1. Confront the superior
2. Inform the superior's boss
3. Ignore the situation

Unfortunately, each option has potential for negative repercussions. Confronting a superior may result in corrective action being taken, or it may result in retribution—or even both. The second alternative may result in similar outcomes. The third alternative compromises one's own ethical standards of behavior. Thus, ethical dilemmas cannot always be expected to be resolved positively. Maintaining one's personal standard of ethics requires action.

Socializing with Other Organization Members

The perception of what constitutes appropriate ethical behavior will affect the actions of fire officers in a social setting. As members of a strong traditional organization, officers are called upon to participate in department-sponsored and other social functions. Some officers feel awkward at these events because the hierarchical structure does not apply in a social setting. The social environment allows a temporary displacement of the official barriers that classify personnel within the organization. It is difficult to identify which specific activities are acceptable and which are not. However, the application of some guidelines for behavior can make these decisions easier.

Rue and Byars Guidelines for
Ethical Conduct in a Social Environment

1. Don't be overly anxious to socialize with subordinates or superiors. Let things take their normal course.
2. Use common sense! Don't do anything while socializing that will cause problems later. The supervisor who does not use common sense when socializing with superiors or subordinates is surely courting trouble.
3. Be yourself! Don't try to put on a false front to impress your chief or other superiors.
4. Don't try to use your rank when socializing with subordinates.
5. Don't make work-related promises to subordinates while socializing.
6. Don't date or become romantically involved with subordinates.[15]

Personal Actions

The personal conduct of fire officers will often influence the actions of other department personnel. Although the hierarchy in the fire service is formed around officer roles and lines of authority, this structure can foster mentoring relationships between the officer and other peers and subordinates. Hence, an officer's actions and conduct serve as examples to be emulated.

The following list outlines leadership characteristics that fire administrators should exhibit as role models and mentors to develop a responsive cadre of personnel.

1. *Show respect for others.* Multicultural municipal agencies are now a reality in many communities, and fire officers need to be sensitive to the opinions and ideas of others, regardless of ethical or cultural background.
2. *Be fair.* Delegate tasks without bias; evaluate personnel on their merits, not on prejudice or hearsay; allow others to speak in open forums.
3. *Accept blame for mistakes.* We all make blunders from time to time, but the key is to admit our faults and keep on going. Most

administrators don't make perfect decisions all the time (they're not expected to!). However, when you make a mistake, don't blame others—accept the consequences, mitigate the impact, and move forward.

4. *Give credit where credit is due.* Be sure that those who have shared ideas and resources with you are recognized for their contributions. Failure to do so demonstrates poor ethical judgment.
5. *Be positive.* The very nature of the fire service tends to place tragedy and sadness at our doorstep. Much of what we hear and observe is negative. Officers need to approach the task of administrating and commanding department personnel with an upbeat, positive attitude. By doing so, others will recognize the value of belonging to and participating in the organization. Being positive and loyal demonstrates a clear understanding of correct ethical behavior.

SUMMARY

The foundation of any ethical public service organization is personal awareness. Executive fire administrators must recognize the need for ethical standards and clearly define the boundaries of ethical conduct within the organization. They should practice ethical principles and encourage other department personnel to do likewise.

The underpinnings of ethics principles are values, integrity, judgment, responsibility, actions, and accountability. Whether derived from codes of ethics, agency rules and regulations, or local and state legislation, ethical standards will drive the organization and play a role in its destiny. Ignoring the need for ethical standards in the fire service conclusively demonstrates a lack of awareness of the future direction of public service and public administration.

ENDNOTES

1. Carol W. Lewis, *The Ethics Challenge in Public Service* (San Francisco, CA: Jossey-Bass, 1991), p. 17.
2. Ibid.

3. Leslie W. Rue and Lloyd L. Byars, *Supervision Key Link to Productivity,* 3rd ed. (Homewood, IL: Irwin, 1990), p. 494.
4. David Adams, "Ethics Infusion," in *Akron Beacon-Journal,* Akron, OH, Feb. 27, p. B1.
5. Rue and Byars, *Supervision Key Link,* p. 400.
6. William A. Ruchan and John W. Newstrom, "How Unethical Are We?," 1975, in Leslie W. Rue and Lloyd L. Byars, *Supervision Key Link to Productivity* (Homewood, IL: Irwin, 1990).
7. Rue and Byars, *Supervision Key Link,* pp. 400–401.
8. Lewis, *The Ethics Challenge,* p. 69.
9. Initial findings from Dr. Nancy Grant, Dr. David H. Hoover, and In Ho Kang, "Fire Officer Ethical Behavior Study" (Summit County, OH: 1993).
10. John A. Granito, "Bye, Bye Birdman," in *Fire Command,* Nov. 1990, pp. 43–44 and Dec. pp. 43–44.
11. Terry L. Cooper, *The Responsible Administrator* (San Francisco: Jossey-Bass, 1990).
12. "Ethics Probe Is All in the Family," in *Akron Beacon Journal,* Akron, OH, March 22, 1992, p. 1.
13. John A. Granito, "A Touchy Situation," in *Fire Command,* Nov. 1990, pp. 43–44 and Dec. pp. 43–44.
14. Cooper, *The Responsible Administrator,* p. 137.
15. Rue and Byars, *Supervision Key Link,* p. 411.

REFERENCES

Adams, David. 1992. "Ethics Infusion" in *Beacon-Journal,* Akron, Ohio, February 27, p. B1.

Akron Beacon Journal. 1992. March 26. p. 17.

Cooper, Terry L. 1990. *The Responsible Administrator.* San Francisco: Jossey-Bass.

Denhardt, Kathryn G. 1988. *The Ethics of Public Service Resolving Moral Dilemmas in Public Service.* Westport, CT: Greenwood Press.

Lewis, Carol W. 1991. *The Ethics Challenge in Public Service.* San Francisco: Jossey-Bass.

NFPA 1021, *Standard for Fire Officer Professional Qualifications,* 1992 ed. Quincy, MA: National Fire Protection Association.

"New Orleans Hosts the 96th" in *NFPA Journal.* 1992. Vol. 86, No. 4. July–August. pp. 74–78.

Rohr, John A. 1989. *Ethics for Bureaucrats: An Essay on Law and Values.* New York: Marcel Dekker, Inc.

Rue, Leslie W., and Lloyd L. Byars. 1990. *Supervision Key Link to Productivity,* 3rd ed. Homewood, IL: Irwin.

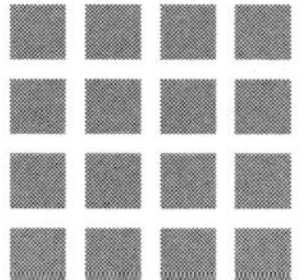

CHAPTER 13

The Future of Fire Service Administration

The fire service is a dynamic organization that has witnessed many changes to its traditional organization and operation. Some of the changes were driven by technology, others were mandated. The fire service will continue to evolve as it meets the challenges of providing services to the public it serves.

Much of the change to date has been driven by federal and state mandates affecting response preparedness and on-scene operations. In addition, fire-service-based innovations such as the incident command system have changed the way in which the fire service is managed. Some of the legislation affecting operations, such as SARA Title III, addresses fire service responsibilities directly. Other laws not written directly for the fire service also have a profound impact.

Equal opportunity and affirmative action legislation mandate equal treatment of women and minorities in hiring and promotion. This legislation has changed the racial and gender mix of fire departments across the nation. Women and minorities are entering the fire service and moving through the officer ranks. It is no longer uncommon to have female officers in command of the scene or to have minorities in chief officer positions. These shifts more accurately reflect the changing demographics of America and provide a convincing indicator that cultural diversity has begun to change the fire service.

Another issue driven by legislation is that of fire fighter safety. The dangerous heroics of the past have given way to safety procedures designed to reduce injuries and deaths. The macho mentality of early fire departments is not in the best interest of either members of the fire service or the citizens of the communities they serve. OSHA regulations mandate a safe work environment, regardless of the task being performed. These mandates have increased the awareness of safety concerns and mitigated the danger of risks taken by fire fighters in response situations. The position of risk manager is becoming more common in fire departments, indicating a commitment to safety on the part of management.

These are only a few of the events that have caused transition in the fire service. As stated earlier, the change is ongoing and the future fire service will be much different from that of the past. Tradition will not be totally replaced; it will continue to have a prominent place in fire service operations. However, the traditions that will last are those that continue to contribute to success.

Serving the Public

For many citizens, becoming part of the fire service has been a goal since early childhood. This desire is sometimes fulfilled by belonging to a local volunteer fire department. Others become career fire fighters, turning their avocation into a full-time vocation. One thing is certain: Women and men in today's fire service are dedicated to their chosen profession. This dedication flourishes in volunteer fire departments as well as in part-paid, combination, and full-time departments.

The fire service must to be recognized as part of the public administration system. In the past, fire chiefs were not considered department managers, by either public administrators or fire fighters. In many instances the best fire fighter became the chief. However, the increasing management responsibilities now placed on fire chiefs is quickly changing this perspective. Fire chiefs must be personnel directors and budget officers as well as leaders and commanders. This combination of technical leadership and administrative ability is part of the path to the future.

Technical Competency

The fire service is one of the most technically trained and competent public service providers in today's communities. Week after week, month after month, fire department members train on specialized apparatus and equipment in preparation for the eventual emergency call. As new techniques are discovered or tools and equipment are developed, department members continue to learn new technical skills. For example, over the past 10 years the concept of pressurizing a building or compartment in order to make it easier for fire service personnel to locate and extinguish structure fires has become a widely accepted fire suppression technique. Equipment manufacturers have responded to this concept by developing new, innovative equipment to enhance the pressurization system. Other new technologies and concepts have also been applied to emergency medical systems, code enforcement inspection procedures, and hazardous materials mitigation. Thus, the technology transfer concept of the 1980s is becoming a reality via the continuous upgrading and expansion of training activities and information sharing.

The type and level of emergency response training currently provided to the fire service is vastly different from training programs of 20 years ago. Terms such as curriculum development, learning outcomes, KSAs, and organizational theory have become the driving force behind many of the current training programs. The "we have always done it that way" approach has been displaced by structured and documented training programs developed by curriculum specialists who validate the materials and the testing instruments. Moreover, in the past 20 years, what was once considered college academic course work has become the curriculum for fire and emergency response training centers. In turn, colleges and universities have developed a higher level of technical curricula in fire protection, fire science, and fire protection engineering. These programs focus on developing new technology, while at the same time training students to conduct research in fire-related topics.

Administrative Responsibility

The U.S. Fire Administration, through the National Fire Academy, recognized the need for officer development beyond technical capabilities and instituted its management science and executive fire officer courses in the 1980s. These original courses have now been upgraded once again to assist fire officers in meeting the increasing level of administrative ability expected.

The Executive Fire Officer (EFO) program continues to build its reputation as a high-quality administrative development program that prepares officers for senior officer leadership positions. Many of the graduates have discovered that learning is, indeed, not only a lifelong but also a career-long process. The annual conferences serve as educational opportunities that help to continuously prepare fire service officers to meet ever-changing administrative challenges.

Similar state and regional officer training and education programs have sprung up throughout the country as the fire service begins to embrace the concept of administrative preparedness. In some instances, these programs are designed around NFPA 1021, *Standard for Professional Fire Officer Qualifications,* while others tend to address local public administrative issues only. Regardless of the type of officer training program in place, the need is acknowledged, and the fire service is changing to address that need.

The next development in this expanding field of fire officer education will be a Master of Fire Administration degree. This has been discussed among leaders of the fire service, of fire education organizations, and of higher education institutions. This degree will become a reality in the near future.

Amelioration

One of the current and future challenges to the fire service is to meet the demands being placed on all aspects of public service. As public administration is challenged, so is the fire service. Thus, it is necessary to respond to the resonant demand to "do more with less." Although we desire to provide the highest recognized level of service possible, we must also begin to understand that there are limitations to the resources available to support our efforts. Thus, the goal of amelioration of fire problems must sometimes replace that of elimination of fire problems. To ameliorate means to improve; it does not mean to eradicate or eliminate. We can never completely eliminate the fire problem; however, through prudent fire prevention efforts, we can make it better. The key issue driving this reality is limited resources.

The financial future for public service providers is uncertain at best. Increased responsibilities and demands for service have been accompanied by a decrease in revenue. The level of revenue available is not likely to increase given the current aversion to tax increases. As the number of high-paying manufacturing jobs continues to decline, and the new wave of service jobs has reduced the level of household income, the amount of discretionary income left to pay for public-sector services has decreased. This is a prominent dilemma in many communities.

There are those who believe that this is a temporary situation due to the slow economic growth of the country. There are others, however, who agree with David Osborne and Ted Gaebler that this is part of a long-term trend based on the unwillingness of the American people to continue to pay more for government services.[1] This has been demonstrated by taxpayers' decisions to forego areas of service or decrease the level of service provided in order to cut back the amount of revenue required. Proposition 13 in California was one of the first manifestations of this philosophy.

Amelioration became part of the focus of public administration discussions in the late 1980s and has continued to be a governing concept in deciding what government will do. For the fire service this involves recognizing that not all lives will be saved and property will be destroyed by fire. *The challenge to the fire service is to lose the fewest lives and the least amount of property, given the resources provided.* Fire departments must make the situation the best they can—not the best of all situations possible. Fire administrators, therefore, are challenged to provide a maximum level of service with diminishing resources.

Fire administrators may have to limit the level of emergency preparedness and response according to the resources available. The safety of the fire fighters must be a primary concern in determining the level of response. If, in fact, a department does not have the equipment or training to effectively control an emergency situation, the response plan must call for securing the area and moving all persons—civilians and fire fighters—to an area of safe refuge.

Many communities cannot afford to be individually prepared for every potential emergency. Fire departments must discover new ways to provide high levels of service while reducing operating costs. One way to accomplish this is through cooperation with surrounding fire departments. Regionalization is the epitome of cooperation, building on activities such as cooperative purchasing, joint training, mutual aid, and automatic responses. Other changes in fire service delivery will include innovations such as cooperative staffing of fire stations, joint purchase and ownership of specialized apparatus, technical specialization among different departments, and consolidation. Fire service officers must improve their administrative skills in order to develop these alternatives successfully.

Regionalization as an Alternative

The notion of combining fire department resources to provide regional emergency response service has recently begun to infiltrate fire department strategic planning efforts. Doing more with less, working smarter, cross-training, and multiple-service delivery agencies are some of the contemporary leadership catch phrases of the 1990s. Fire departments are beginning to experience the reallocation of resources due to prioritization efforts by elected and public officials. The question is, *Can fire departments really be prepared to handle all potential emergencies that arise?* The immediate answer is that the response must be commensurate with the resources available.

Even with changing technology and modern command and administrative systems in place, the fire service of today must make some difficult and reasoned choices as it prepares for emergency response. The level and type of emergency response in a community must match the level of response capability present in the community. Once the hazard is analyzed and the level at which the community is prepared to respond is determined, the emergency response is then designed accordingly. This may require some difficult decisions for fire service officers. When the resources are not available for comprehensive response, the local response must be restricted to immediate mitigation and summoning assistance. To do other than this would place the members of the fire department in danger beyond that required during their normal duties. Safety of fire fighters must be of paramount concern as response decisions are made. Response must be kept in line with available resources. As revenue continues to erode due to changes in public priorities and a slowed economy, the fire service will be challenged to work smarter, do more with less, and cut costs through cooperation. As part of the public service, it will have to address the same challenges.

Although some fire service traditions will remain in some form, many of the artificial barriers that have been erected over the past 50 years must be removed. For instance, career, volunteer, and on-call personnel will work together in a strong, effective emergency service provider system. We must recognize that placing personnel safety above traditional beliefs and values is fundamental to the future growth of the fire service. A sense of pride is important to the success of any organization, but these feelings need to be kept in perspective and not be allowed to overshadow the community needs. To modify some of the differences that have caused divisions among the various elements of the fire service, a new spirit of cooperation must be embraced.

The place to start is the fire department mission statement. Read it and compare your mission to the mission statements of other fire departments in the region. Remarkably, you will discover a common bond or thread exists that transcends the boundaries of all fire departments and addresses the issue of service delivery at the highest level. Now ask the question, What are the short-term, long-term, and strategic goals of the organization? Do these goals provide the direction necessary to achieve the mission of the department? Then, match the department's goals to those of other fire departments in the region. Are the goals consistent across geographic boundaries and do they hold the needs of local citizens at the highest level? Establishing the area of commonality is the first step toward regional cooperation.

The next step is training. In addition to the standard review and upgrading of fire department training programs, increased attention must be paid to cooperative, regional training. Fire departments need to begin to train together. This reduces individual department costs and better prepares the fire fighters to work together during an emergency incident. The incident command system has laid the foundation for common understanding of terminology and operations, facilitating cooperative training and response.

Not only does a community's fire department need to cooperate with adjacent fire departments, it must also cooperate with other emergency response agencies. This means that working arrangements must be established with the local and state law enforcement and emergency management agencies as well as health departments and other municipal service providers in the area served. The fire department must work effectively with these agencies in the event of a major catastrophe, as well as during specific types of emergency situations. As a result, cooperative working arrangements must be developed ahead of time.

The fire service does not exist in isolation. It is, in fact, one of the components of the larger arena of public service. Therefore, the challenge of the fire service is to meet the demands of the public service within the context of public administration operations.

The Future Public Administrator

The successful future fire chief cannot be a caretaker. The leaders of the future must be innovative. "Entrepreneurial leaders" are touted as the public administrators who will achieve success in the future. These individuals develop innovative ways of accomplishing their mission. They do not rely on the past, but instead examine creative approaches to accomplishing goals in very different ways.

For example, as budgets continue to decline, the fire chief of the future must begin to challenge the officer staff to identify ways of cutting costs and generating new revenue within the department. Why not ask fire officers to identify fifteen innovative ways to generate revenue, in addition to standard permit fees? Department personnel can be an important source of new ideas. Even though they may not be senior fire officers, many fire fighters are achieving a higher level of education and have ideas that can be adapted to the fire service. Innovative ideas cannot be ignored. The fire service must not only catch up with what is going on in the public sector and in the community, but must take the lead in responding to these changes with innovative leadership and service delivery.

Fire Department Customers

The citizens of the community are the customers of the local fire department. Without citizens, there is no need for the fire department. Fire service officers must respond to this reality and recognize that they must keep their customers satisfied. "Service quality is a central issue in America today. The United States may be on the verge of taking the same beating in services that it has already endured in manufacturing," and the "principal culprit is seen as mediocre service quality."[2] This is as vital for the fire service as it is for any other public service department or agency.

Firc administrators must bc conscious of thc importancc of customer service and need to encourage and support customer service within the departmcnt's ranks.

> *It is not that American service providers are slovenly or lazy. We have found absolutely no evidence in our research of a poor work ethic in America. Rather, what happens is that new, high-energy service employees do not receive the organizational support and inspiration they need to sustain them through the inevitable travail of service work. And without the support of good leaders, good teammates, clear direction, (and) consistent signals . . . many of these service providers lose energy and effectiveness over time even as they increase in competence. They lose the will to serve.*[3]

Thus, the challenge for fire officers is to motivate and provide support for good performance, to encourage innovation, and to provide a consistently high-quality service in a changing society.

ENDNOTES

1. David Osborne and Ted Gaebler, *Reinventing Government: How the Entrepreneurial Spirit Is Transforming the Public Sector* (New York: Addison-Wesley Publishing Co., 1992).

2. Valerie A. Zeithaml, A. Parasuraman, and Leonard L. Berry, *Delivering Quality Service: Balancing Customer Perceptions and Expectations* (New York: The Free Press, 1990). p. 139.
3. Ibid., pp. 141–142.

REFERENCES

Kennedy, Paul. 1993. *Preparing for the Twenty-First Century.* London: Harper Collins.

Johnson, William, and Arnold Packer, et al. 1987. *Workforce 2000: Work and Workers for the 21st Century.* Indianapolis, IN: Hudson Institute.

Index